Narratives of Low-Carbon Transitions

T0187711

This book examines the uncertainties underlying various strategies for a low-carbon future. Most prominently, such strategies relate to transitions in the energy sector, on both the supply and the demand side. At the same time they interact with other sectors, such as industrial production, transport, and building, and ultimately require new behaviour patterns at household and individual levels. Currently, much research is available on the effectiveness of these strategies but, in order to successfully implement comprehensive transition pathways, it is crucial not only to understand the benefits but also the risks.

Filling this gap, this volume provides an interdisciplinary, conceptual framework to assess risks and uncertainties associated with low-carbon policies and applies this consistently across 11 country cases from around the world, illustrating alternative transition pathways in various contexts. The cases are presented as narratives, drawing on stakeholder-driven research efforts. They showcase diverse empirical evidence reflecting the complex challenges to and potential negative consequences of such pathways. Together, they enable the reader to draw valuable lessons on the risks and uncertainties associated with choosing the envisaged transition pathways, as well as ways to manage the implementation of these pathways and ultimately enable sustainable and lasting social and environmental effects.

This book will be of great interest to students, scholars, and practitioners of environmental and energy policy, low-carbon transitions, renewable energy technologies, climate change action, and sustainability in general.

Susanne Hanger-Kopp is a researcher at the Institute for Environmental Decisions at ETH Zürich, Switzerland, and at the Risk and Resilience Program of the International Institute for Applied Systems Analysis in Laxenburg, Austria.

Jenny Lieu is a research fellow at the University of Sussex, UK, and guest researcher at the Transdisciplinary Lap of the Department of Environmental Systems Science at ETH Zürich, Switzerland.

Alexandros Nikas is a researcher at the Management & Decision Support Systems Laboratory of the School of Electrical and Computer Engineering, National Technical University of Athens, Greece.

Routledge Studies in Energy Transitions

Considerable interest exists today in energy transitions. Whether one looks at diverse efforts to decarbonise, or strategies to improve the access levels, security and innovation in energy systems, one finds that change in energy systems is a prime priority.

Routledge Studies in Energy Transitions aims to advance the thinking which underlies these efforts. The series connects distinct lines of inquiry from planning and policy, engineering and the natural sciences, history of technology, STS, and management. In doing so, it provides primary references that function like a set of international, technical meetings. Single and co-authored monographs are welcome, as well as edited volumes relating to themes, like resilience and system risk.

Series Editor: Dr. Kathleen Araújo, *Boise State University and Energy Policy Institute, Center for Advanced Energy Studies (US)*

Series Advisory Board
Morgan Bazilian, *Colorado School of Mines (US)*
Thomas Birkland, *North Carolina State University (US)*
Aleh Cherp, *Central European University (CEU, Budapest) and Lund University (Sweden)*
Mohamed El-Ashry, *UN Foundation*
Jose Goldemberg, *Universidade de Sao Paolo (Brasil) and UN Development Program, World Energy Assessment*
Michael Howlett, *Simon Fraser University (Canada)*
Jon Ingimarsson, *Landsvirkjun, National Power Company (Iceland)*
Michael Jefferson, *ESCP Europe Business School*
Jessica Jewell, *IIASA (Austria)*
Florian Kern, *Institut für Ökologische Wirtschaftsforschung (Germany)*
Derk Loorbach, *DRIFT (Netherlands)*
Jochen Markard, *ETH (Switzerland)*
Nabojsa Nakicenovic, *IIASA (Austria)*
Martin Pasqualetti, *Arizona State University, School of Geographical Sciences and Urban Planning (US)*
Mark Radka, *UN Environment Programme, Energy, Climate, and Technology*
Rob Raven, *Utrecht University (Netherlands)*
Roberto Schaeffer, *Universidade Federal do Rio de Janeiro, Energy Planning Program, COPPE (Brasil)*
Miranda Schreurs, *Technische Universität München, Bavarian School of Public Policy (Germany)*
Vaclav Smil, *University of Manitoba and Royal Society of Canada (Canada)*
Benjamin Sovacool, *Science Policy Research Unit (SPRU), University of Sussex (UK)*

Narratives of Low-Carbon Transitions

Understanding Risks and Uncertainties

Edited by Susanne Hanger-Kopp, Jenny Lieu, and Alexandros Nikas

Routledge
Taylor & Francis Group
LONDON AND NEW YORK

earthscan
from Routledge

First published 2019
by Routledge
2 Park Square, Milton Park, Abingdon, Oxon OX14 4RN

and by Routledge
52 Vanderbilt Avenue, New York, NY 10017

First issued in paperback 2020

Routledge is an imprint of the Taylor & Francis Group, an informa business

© 2019 selection and editorial matter, Susanne Hanger-Kopp, Jenny Lieu, and Alexandros Nikas; individual chapters, the contributors

The right of Susanne Hanger-Kopp, Jenny Lieu, and Alexandros Nikas to be identified as the authors of the editorial matter, and of the authors for their individual chapters, has been asserted in accordance with sections 77 and 78 of the Copyright, Designs and Patents Act 1988.

The Open Access version of this book, available at www.taylorfrancis. com, has been made available under a Creative Commons Attribution-Non Commercial-No Derivatives 4.0 license.

Trademark notice: Product or corporate names may be trademarks or registered trademarks, and are used only for identification and explanation without intent to infringe.

British Library Cataloguing-in-Publication Data
A catalogue record for this book is available from the British Library

Library of Congress Cataloging-in-Publication Data
Names: Hanger-Kopp, Susanne, editor. | Lieu, Jenny, editor. | Nikas, Alexandros, 1989- editor.
Title: Narratives of low-carbon transitions : understanding risks and uncertainties / edited by Susanne Hanger-Kopp, Jenny Lieu and Alexandros Nikas.
Description: Milton Park, Abingdon, Oxon ; New York, NY : Routledge, 2019. | Series: Routledge studies in energy transitions | Includes bibliographical references and index.
Identifiers: LCCN 2018052683 (print) | LCCN 2018055200 (ebook) | ISBN 9780429458781 (Master) | ISBN 9781138311589 (hbk) | ISBN 9780429458781 (ebk)
Subjects: LCSH: Renewable energy sources. | Energy conservation. | Carbon dioxide mitigation.
Classification: LCC TJ808 (ebook) | LCC TJ808 .N37 2019 (print) | DDC 621.042–dc23
LC record available at https://lccn.loc.gov/2018052683

ISBN 13: 978-0-367-66071-0 (pbk)
ISBN 13: 978-1-138-31158-9 (hbk)

Typeset in Goudy
by Wearset Ltd, Boldon, Tyne and Wear

Contents

Figures

Tables

Contributors

Rocio Alvarez-Tinoco is Research Fellow at SPRU (Science Policy Research Unit), University of Sussex. She got her MSc and PhD degree in Science and Technology Policy Studies at SPRU. Her current research focuses on risks and uncertainties in low-carbon mitigation pathways for nuclear and renewables. She has research experience in capabilities development, institutional change, and has worked in several Latin American countries and on R&D and technology transfer in firms.

Apostolos Arsenopoulos is a PhD candidate at the Management & Decision Support Systems Laboratory of the School of Electrical and Computer Engineering, National Technical University of Athens. He has been intensively involved in power systems issues and energy–economy system modelling; his interests mainly lie in climate–energy–economy modelling, operational research, and extended portfolio analysis.

Iñaki Arto is a senior researcher at BC3 (Basque Centre for Climate Change). He has a PhD in Economics and his research focuses on environmental–economic modelling (Integrated Assessment, Computable General Equilibrium, Dynamic Econometric Input–Output), social metabolism, and clean production.

Gabriel Bachner is a post-doctorate researcher and member of the EconClim Research Group at the Wegener Center for Climate and Global Change (University of Graz). His fields of research focus on the economics of climate change impacts, adaptation, and mitigation, for the latter particularly in the energy and transport sectors. He has extensive experience in macroeconomic general equilibrium modelling and also stakeholder involvement.

Krisztina de Bruyn-Szendrei graduated in chemistry at the University of Szeged, Hungary. She obtained her PhD in natural sciences at the University of Groningen, in the Netherlands, based on research on developing novel renewable solar cells. Afterwards, she joined JIN Climate and Sustainability, the Netherlands, where she worked for four-and-a-half years as a researcher on topics in the framework of renewable energy.

Rodrigo Cerda received his PhD from the University of Chicago in 2003. He is an associate professor at the Pontificia Universidad Católica de Chile, and since 2014 also co-director of the Centro LatinoAmericano de Políticas Sociales y Económicas of the same University. His main research interests are macroeconomics and development.

Ying Chen is a senior research fellow at the Institute of Urban and Environmental Studies (IUE), Chinese Academy of Social Sciences (CASS), Deputy Director of the CASS Research Center for Sustainable Development (RCSD), and Professor at CASS Graduate School. Since 2013 she has been a Chief Researcher Fellow for Innovative Program in IUE, CASS. She worked as a lead author for the Intergovernmental Panel on Climate Change (IPCC), the Fifth Assessment Report (AR5), and Sixth Assessment Report (AR6). She is a vice chairman of the Chinese National Committee for Future Earth (CNC-FE).

Ed Dearnley works at the Science Policy Research Unit (SPRU) of the University of Sussex. He has 18 years' experience of project management, public policy, and business analysis, mainly in the environmental sector (air quality, climate change, and energy efficiency). Prior to joining SPRU, he worked as a policy officer for the UK non-governmental organisation Environmental Protection UK, before starting his own air quality and climate policy consultancy.

Tahia Devisscher is a research associate at the Stockholm Environment Institute, Oxford Centre, UK, and a postdoctoral research fellow at the University of British Columbia, Canada. Since 2008, she has worked on development and the environment, mainly on climate change adaptation. Tahia applies an interdisciplinary approach to study environmental issues and build social–ecological resilience. She has consulted for the World Bank, UNEP, and UNDP, and engaged in 15 international research projects with partners across the globe. She obtained her PhD in 2016 from the University of Oxford.

Haris Doukas is an assistant professor at the School of Electrical and Computer Engineering of the National Technical University of Athens. His scientific and research expertise includes the development of models and decision support systems for energy and environmental policy and management. He has participated, as a scientific coordinator/project manager/main expert, in the design of energy policies and programmes promoting renewables, energy efficiency, and rational use of energy at a local, national, and European level.

Cecilia (Cece) Fitzpatrick, also known as White Bear Woman, is an Aboriginal woman leader and elder born to Phillip (former chief) and Victoria McDonald in Fort McKay First Nation, Alberta. Cecilia grew up travelling through the trap lines with her family and learned how to live off the land. She has experience in the oil and gas industry and governance of Fort McKay First

Nation. She is a passionate environmental advocate for Indigenous communities in Canada and has the role of providing companionship and care for other senior Elders.

Aikaterini Forouli is a PhD student at the Management & Decision Support Systems Laboratory of the School of Electrical and Computer Engineering, the National Technical University of Athens. Her expertise lies in portfolio analysis and robustness techniques for the support of decision making in environmental policy issues and energy planning. She has further interest in stakeholder-driven multi-criteria analysis methods and climate–economy modelling frameworks for climate change simulation and mitigation/adaptation policy evaluation.

Wytze van der Gaast has worked for JIN Climate and Sustainability since 1995 on research and policy advice on emissions trading, technology transfer, and low-emission transitions. He has been involved in 16 EU-funded projects and has been regular adviser to UN bodies on climate policy topics.

Xaquín García-Muros is a postdoctoral researcher at the Basque Centre for Climate Change (BC3) and holds a PhD in Economics from the University of the Basque Country (2017). His main research interest is to create and spread knowledge in the field of energy, climate change and public economics, and the distributional impacts of climate change mitigation policies.

Nikolaos Gkonis is a senior energy efficiency expert and a PhD student at the Management & Decision Support Systems Laboratory of the School of Electrical and Computer Engineering, the National Technical University of Athens. He has a deep knowledge of energy efficiency and energy services issues, as well as a remarkable track record of projects both in technical and policy framework. As a consultant to the Greek Ministry of Environment and Energy, he is a member of the working group for policies, measures, and models for the preparation of the Greek National Energy and Climate Plan.

Luis E. Gonzales Carrasco worked in the government of Bolivia in the Social and Economic Policy Analysis Unit (UDAPE) (2005–2008), was adviser to the Ministry of Finance of Chile (2012–2014), and is currently an Associate researcher at the Latin American Center for Economic Policies and Social CLAPES UC of the Pontificia Universidad Católica de Chile. He has a Masters degree in Public Policy from Harris School of Public Policy at the University of Chicago in the United States.

Mikel González-Eguino is a senior researcher at BC3 (Basque Centre for Climate Change) and has a PhD in Economics (University of the Basque Country, 2006). His main interests lie in learning how the integration of social sciences and natural sciences can support decision making in the transition to a low-carbon, sustainable economy.

Susanne Hanger-Kopp is a research scholar in the Risk and Resilience Program at the International Institute for Applied Systems Analysis. She received her doctorate from ETH Zürich, where she works and teaches in the Climate Policy Group. Her main expertise and interests are in the areas of climate policy, decision making, and governance, with a focus on climate risk management and risk perception using soft systems approaches.

Cynthia Ismail is a researcher at PT Sustainability & Resilience (su-re.co), Bali, Indonesia, since 2016 and has been involved in a number of collaborative projects with EC, UNIDO, UNDP, and ADB. With a main focus on climate change adaptation and mitigation, she contributed to bioenergy development in East Java and Bali. She has worked on climate change scopes since 2012 and obtained an MSc degree in Management and Engineering of Environment and Energy from KTH Royal Institute of Technology.

Francis X. Johnson is a senior research fellow at the Stockholm office of SEI. He conducts interdisciplinary energy/climate analyses, capacity building and research, focusing especially on biomass energy in developing countries. He has over 20 years of experience in economic and environmental analysis of bioenergy, climate mitigation, and energy efficiency. Prior to joining SEI, he was a senior research associate in the Energy Analysis programme at Lawrence Berkeley National Laboratory, USA. He has served as an adviser or expert for international initiatives run by UNIDO, FAO, the European Commission, and the Environment Committee of the European Parliament.

Oliver W. Johnson is a senior research fellow and Energy Cluster Leader at SEI's Asia Centre. He has over ten years' professional experience in energy, climate change, and international development. He has worked in Africa and Asia, with extensive experience in Zambia, Tanzania, Ghana, Kenya, Ethiopia, Rwanda, Cambodia, and India. Oliver has been employed by a diverse range of organisations, including international organisations, national governments, community-based NGOs, development policy think tanks, and academic institutions. His areas of work include energy access, energy sector governance, low-carbon innovation, and climate change.

Eleni Kanellou is a PhD student at the Management & Decision Support Systems Laboratory of the School of Electrical and Computer Engineering, the National Technical University of Athens. She has experience in stakeholder engagement and decision support systems for supporting energy policy and planning. Her research interests include climate change and policy, sustainable business, and environmental planning.

Charikleia Karakosta is a chemical engineer holding a PhD in Decision Support Systems Promoting the Effective Technology Transfer within the Frame of Climate Change and an MSc in Energy Production and Management from NTUA. She has published more than 145 scientific publications in prestigious international journals and books with numerous citations. For her work she has received the ECOPOLIS Award 2016 for young scientist's research and awards by the Onassis Foundation, the State Scholarship Foundation, and the NTUA.

Jenny Lieu is a research fellow at SPRU (Science Policy Research Unit) in the University of Sussex. She is currently the Co-Principal Investigator for TRANSrisk, a Horizon 2020 project that evaluates risks and uncertainties of low-carbon transition pathways at national, regional, and global levels. She has a diverse research interest in energy and climate policy and innovation studies. Jenny has provided consultancy services, producing strategic recommendations for energy and environmental infrastructure firms and conducting policy analysis for international institutions.

Gordon MacKerron is Professor of Science and Technology Policy at SPRU (Science Policy Research Unit), University of Sussex, having been Director of SPRU 2008–2013. He is an economist working mainly in energy, environmental economics, and energy policy, specialising in electricity and nuclear power, climate change mitigation, and energy security.

Jakob Mayer MSc is member of the EconClim Research Group at the Wegener Center for Climate and Global Change and a PhD student in Economics & Social Sciences at the University of Graz. His research covers climate change economics, particularly regarding the transition to low-carbon or carbon-free socio-economic systems. He is experienced in quantitative (computable general equilibrium modelling) and qualitative methods (stakeholder interaction).

Alexandros Nikas is a PhD student at the Management & Decision Support Systems Laboratory of the School of Electrical and Computer Engineering, the National Technical University of Athens. He has long-standing experience in designing and utilising scientific processes and systems in support of energy and climate policymaking, as well as of bridging the climate–policy interface. His interests lie in climate–economy modelling, operational research, and stakeholder participatory frameworks.

Mbeo Ogeya is a research fellow at Stockholm Environment Institute in the Energy and Climate Change Programme. He has 11 years' research experience in renewable energy technologies. He is a Long-range Energy Alternative Planning tool expert with skills in energy systems modelling, engineering design, solar photovoltaic, bioenergy, and energy efficiency. His research area seeks to address energy trends and associated flows across all levels of production, distribution, and utilisation, from micro to macro systems for effective policy formulation.

Aikaterini Papapostolou is a PhD candidate at the Management & Decision Support Systems Laboratory of the School of Electrical and Computer Engineering, the National Technical University of Athens. Her scientific experience falls into the areas of energy (environmental policy and planning and decision support methods). Her research focuses on the assessment of renewable energy cooperation between Europe and its neighbouring countries using multiple-criteria decision support methods.

Cristina Pizarro-Irizar is an assistant professor at the University of the Basque Country and associate researcher fellow at BC3. She holds a PhD in Economics from the University of the Basque Country. Her research focuses on the economics of renewable energy, mainly in the analysis of the policies aimed at renewable energy support in Spain.

Mariana Silaen has worked for PT Sustainability & Resilience (su-re.co), Bali, Indonesia, in the scope of research and hands-on activities related to climate change and renewable energy in rural areas with interlinkage to wider impacts with various stakeholders. She holds a bachelor's degree in meteorology from Bandung Institute of Technology (ITB) and a Master's degree in International Development from the University of Manchester.

Eleftherios Siskos is a chemical engineer and currently a PhD candidate in the Decision Support Systems Laboratory of NTUA. His research interests fall into the areas of multi-criteria decision analysis, multi-objective programming, and robustness analysis of decision support models. The optimisation problems he mostly focuses on concern energy planning and environmental management towards sustainable development.

Lei Song is Associate Professor of Urban Sustainable Development at the China Executive Leadership Academy, Pudong, China. She is also a post doctor in the Institute of Urban Sustainable Development and Environmental Studies, CASS. Now her research field focuses on low-carbon economics, climate change, and ecosystem services.

Alevgul H. Sorman is a postdoctoral researcher at the Basque Centre for Climate Change (BC3) with a PhD (2011) from the Institute of Environmental Science and Technology (ICTA) of the Autonomous University of Barcelona (UAB). Her work focuses on interfacing energy and societal metabolism with justice concerns on changing energy frontiers.

Eise Spijker is a senior adviser and researcher on market and innovation system analysis, low-emission transition pathways, bioenergy, sustainable agriculture, emissions trading and policy interactions. He has participated in several EU-funded research projects on these topics and is also a regular adviser to the Netherlands government.

Karl Steininger is Professor at the Department of Economics at the University of Graz and Head of Economics of the Climate and Global Change Research Group (EconClim) at the Wegener Center. He specialised in environmental and ecological economics and in international trade, and has led national and international interdisciplinary research projects for more than two decades (on behalf of the World Bank, the OECD, and various national research funding agencies). He has long-lasting research experience in climate and energy research, in particular in empirical macroeconomic and climate policy quantitative modelling.

Michele Stua is Research Fellow at SPRU (Science and Policy Research Unit), University of Sussex, where he has spent most of his time as an academic and researcher. His major studies aimed at facilitating the trade of mitigation-related assets worldwide. His findings led to the definition of policy proposals introduced in multiple international contexts. He holds a relevant track of publications ranging from global climate change governance topics to the effects of Brexit on the UK's climate and energy strategies.

Takeshi Takama has been an international expert on climate change, environment, and energy for international and bilateral agencies for more than 15 years, including ADB, IRENA, JICA, GIZ, and UN agencies. He is also an associate for the Stockholm Environment Institute, CEO of PT Sustainability & Resilience (su-re.co), and a professor at Udayana University. His expertise includes climate change vulnerability and adaptation, renewable energy with particular scope of bioenergy, green economy, and food security. He completed his PhD from the University of Oxford in 2005.

Richard Taylor is a senior researcher at the Stockholm Environment Institute, Oxford Centre, UK. His work addresses decision-making, behavioural, and learning processes related to environment, climate change, and sustainable development. A focus is on using participatory methodologies with stakeholders, and provision of decision support processes and tools. He also develops agent-based social simulation (ABSS) models and data visualisations in environmental management and development contexts. He obtained his PhD in 2004 at MMU Business School.

Syamsidar Thamrin has been working for nearly 20 years with the Indonesian National Development Planning Agency (Bappenas), currently being a senior planner for energy and climate change. She has been involved in various climate change policies in Indonesia, government procurement policy development, and the energy sector. She was also assigned as the Head of Secretariat of ICCTF in 2009–2015. She completed her bachelor's degree in Electrical Engineering from Bandung Institute Technology (ITB) and a Master's degree in Business Administration from the Syracuse University.

Andreas Tuerk MSc, MBA is a senior scientist with an extensive international background. The chemist and economist works at both Joanneum Research and the Wegener Center for Climate and Global Change (University of Graz). He has more than ten years' expertise in international and national energy and climate policy, emissions trading (e.g. the EU Emissions Trading Scheme, international emissions trading schemes), the Kyoto Protocol and subsequent international regimes/agreements, energy and environmental policy interaction analyses, smart grid sand storage, and smart cities, as well as regulatory issues related to the design of electricity markets.

Luis D. Virla is Research Fellow of the Science Policy Research Unit (SPRU) of the University of Sussex. His main research interests comprise the development of catalytic technologies for CO_2 capture and utilisation, as well as studying energy transition pathways towards low-carbon technologies. Luis holds a PhD from the University of Calgary and a MSc/BSc in Chemical Engineering from the University of Zulia.

Oscar van Vliet received his Master's degrees and a PhD in environmental science from Utrecht University. His main research focus has been on the question of why we are not all switching to renewable energy. He works on this with a wide variety of methods including systems modelling, techno-economic assessment, public surveys, and stakeholder consultations. Oscar has also worked in private consultancy and at the International Institute for Applied Systems Analysis. He is currently a lecturer with the Climate Policy Group at ETH Zürich.

Brigitte Wolkinger, Mag., BEd is a member of the EconClim Research Group at the Wegener Center for Climate and Global Change (University of Graz). Her main fields of research cover topics of environmental economics and econometrics, i.e. reduction of greenhouse gas emissions and transport economics. She has gained extensive expertise in moderation of stakeholder processes and stakeholder analysis.

Hannah Wanjiru is a research associate at SEI Africa, with over eight years' experience working on energy and environment issues in the region. Her research work cuts across policy development, biomass value chain, gender mainstreaming, technology adoption, greenhouse gases inventory, and standards. Prior to joining SEI, Hannah worked as an energy project officer at Practical Action's East Africa Regional Office, Nairobi, where she contributed to the design and delivery of projects on renewable energy assessment, biomass technologies, energy markets, cookstoves, policy advocacy, and mini-grids.

Yudiandra Yuwono graduated as an environmental engineer from Bandung Institute of Technology (ITB) in 2014. Since then he has worked for private sectors and NGOs, gaining experience in different industries including mine water management, oil waste, soil recovery, and sustainable technology distribution. He has worked in various places with international agencies such as EC, UNDP, and IRENA.

Preface and acknowledgements

In the face of climate change, resource depletion, environmental pollution, and other global challenges, attempts to transition to a low-carbon society are ongoing across the world. International negotiations and agreements, national policy strategies, and choices at firm and household levels all contribute to the development and uptake of technological innovations, and are crucial for any sustainable, low-emission future. The interplay of these multi-level activities, choices, enabling factors, and potential negative side effects are complex and difficult to disentangle. One important step towards this is to explore national, regional, and local transition efforts. Here we can consider both specific technologies and enabling policies, and the social context in which they are supposed to function.

This book illustrates some of the risks and uncertainties associated with low-carbon transitions, which are tightly linked to changes and innovations in the sector, as they are currently happening, planned, or envisaged in different national contexts. We experiment with narratives to present our results, aiming to make our research accessible to a wider audience that includes, but goes beyond, our academic peers. We firmly believe that it is important for scientists and researchers to communicate their research findings in multiple ways, channels, and formats. We aim to reach as broad an audience as possible in order to inform science, policymaking, and other relevant social processes.

Our format, which we employed across all case study chapters, allows us to go beyond the limitations of scenario building based on models. Formalised model frameworks cannot adequately represent the many non-quantifiable aspects of policies and their associated risks and uncertainties. Our format allows us to include the perceptions and, more importantly, the expertise of a wide range of stakeholders in energy transition development, in an effort to make recommendations that are socially acceptable and robust.

Our work is interdisciplinary throughout and transdisciplinary in many instances. The lengthy process needed to move from merely talking about interdisciplinary work to actually doing interdisciplinary work is indicative of the time it will take to master the much more recent idea of transdisciplinary work, where the focus lies on co-development with non-academic stakeholders. We hope this book will help us move closer to real transdisciplinary knowledge production.

We would like to thank all of the stakeholders who engaged with our case studies. Without their participation, patience, and engagement we would not have been able to collect, assess, and disseminate this rich body of knowledge.

All the chapters in this book have been reviewed by one to three experts in the case study fields, in a transparent manner, i.e. both authors and reviewers were aware of each other's role. We believe this is a good alternative to a double-blind peer-review process, as the critical yet respectful and constructive feedback has been extremely useful in order to improve our work.

For their efforts, we would like to thank the following reviewers: Jialing Lu (University of Sussex), Thomas Schinko (Risk and Resilience Program, International Institute for Applied Systems Analysis), Milan Elkerbout (Centre for European Policy Studies), Stefan Bössner, Tadeusz Skoczkowski (Department of Rational Use Of Energy, Warsaw University of Technology), Stavros Vlachos (EMICERT), Theocharis Pitsilis (Envirometrics), Mike Middleton (Energy Technologies Institute), Joan David Tábara (Institute of Environmental Sciences and Technology of the Autonomous University of Barcelona), Subhasis Ray (Xavier Institute of Management, Xavier University), Evans Kituyi (International Development Research Centre, Canada), and Michael Stauffacher (Transdisciplinarity Lab, Department of Environmental System Science, ETH Zürich). We also like to thank those reviewers who, for different reasons, remain anonymous.

Furthermore, we extend our gratitude to those who have supported the engagement processes, provided advice, and given valuable feedback: Phil Johnstone (University of Sussex) and Gert-Jan Kok (JIN Climate and Sustainability).

The remarkable and reliable editorial support of Ed Dearnley and Ellie Leftley (University of Sussex) added immeasurably to the quality of each chapter and the overall organisation of the book.

The book, and every individual chapter, were the result of several years of research work under the project 'Transitions pathways and risk analysis for climate change mitigation and adaptation strategies – TRANSrisk', which received funding from the European Union's Horizon 2020 Research and Innovation Programme under grant agreement number 642260.

The chapter on Austria also received funding from the Austrian national bank, Jubiläumsfonds, under project grant number 16282 (project 'EFFECT'). The Swiss chapter benefited from funding by the National Research Programme 'Energy Turnaround' (NRP 70) of the Swiss National Science Foundation (SNSF), under grant number 1153974 (project 'New Risks').

The sole responsibility for the content of each case study chapter, and of the other components of the book, lies with the authors and editors respectively. The book does not necessarily reflect the opinion of the European Commission or other funding institutions. Finally, many risks and uncertainties in this book represent stakeholder perspectives, while in several instances stakeholders have co-created the pathways at the heart of each narrative. Therefore, the narratives do not necessarily reflect the views of the organisations the stakeholders are affiliated with or the authors' views.

Susanne Hanger-Kopp, Jenny Lieu, and Alexandros Nikas

Part I
Setting the stage

1 Introduction

Susanne Hanger-Kopp, Jenny Lieu, and Alexandros Nikas

In this book, we tell stories of low-carbon transitions – which are almost always also energy transitions – as a means to illustrate associated risks and uncertainties. Stories or narratives are descriptive, non-technical accounts of work that goes beyond single academic disciplines, highlighting the perspectives of the stakeholder groups involved. They allow for a thorough qualitative account of case-specific detail with respect to policy and governance processes and stakeholder perceptions, which are important for understanding both transitions and the risks and uncertainties associated with and inherent in these transitions. They allow for a comprehensive inter- and transdisciplinary view of risks and uncertainties, which is critical for enhancing risk governance. Finally, they allow for a writing style accessible to a wider, non-academic audience, which is particularly important for informing policymaking processes.[1]

Most recently, storytelling found its way into international policy processes. The Talanoa Dialogue, introduced at the 21st Conference of the Parties (UNFCCC, 2015), aims at sharing stories and building empathy and trust. Talanoa is a tradition of inclusive, participatory, and transparent dialogue from the South Pacific. Such a process enables participants to advance knowledge through common understanding and prospectively contributes to better collective decision making that advances the global decarbonisation and adaptation agenda (UNFCCC, 2016). Where are we? Where do we want to go? And how do we get there? These are the key questions structuring the dialogue, essentially leading to narratives about low-carbon transitions.

These transitions are essential to achieve decarbonisation and combat climate change but come with a host of associated risks and uncertainties. As a key element in each of the narratives presented in this book, transition pathways describe what the world should look like over time for realising a common goal. Said descriptions can include, but are not limited to, technological choices, institutional set-ups, and drivers for behavioural and infrastructure changes. Aiming at a low-carbon future, such pathways include global agreements and climate targets, national strategies and emission targets, as well as concrete policy instruments to implement them. These in turn rely on resource allocations and investments, as well as behavioural change at both organisational and individual levels. Such pathways are not forecasts but represent deliberately defined cases of past, present, and future system development. They allow useful exploration of, and learning from, states of the world, given that their narratives are salient, credible, and legitimate (Cash *et al*. 2003, Rounsevell and Metzger, 2010).

The narratives in this book have been developed through the work of the EU Horizon 2020-funded TRANSrisk project. They refer to national and subnational levels but are considered in the broader context of global climate change and decarbonisation governance. Globally, the Paris Agreement constitutes the most important overarching aim for low-carbon transitions. It is a high-level legal framework attempting to pave the way for global low-carbon and resilient development. The agreement's binding nature is weakened by the absence of sanctions, its non-punitive, non-adversarial design, and the lack of clear implementation policies. Its key elements are the organisation of the contributions in terms of emission reductions by the parties; the relationships and transactions between developed, developing, and least developed countries; and funding mechanisms to realise actions. The Paris Agreement addresses both mitigation and adaptation concerns – with respective implications for and explicit focus on finance, technology transfer, and capacity building – and most recently also considerations for loss and damage. With respect to mitigation, the most surprising outcome of the 21st Conference of the Parties (COP) was the introduction of the new, highly ambitious mitigation target aiming to limit temperature rise to below 1.5°C. Although the target of 2°C remains part of the agreement as the upper limit temperature target, this implies a radical redefinition of low-carbon pathways as discussed, analysed, and assessed up until 2015.

Parties' contributions to the agreement are embodied in their nationally determined contributions (NDCs). Based on the agreement and the newly introduced stocktaking mechanism, each party prepares, communicates, and maintains their NDCs autonomously, taking full responsibility for their implementation while respecting the principles of environmental integrity, transparency, accuracy, completeness, comparability, and consistency. The Paris Agreement, however, does not provide procedures and methodologies for the determination of NDCs. This means that, ultimately, any specific action towards the targets of the Paris Agreement is the responsibility of individual countries.

The EU as a unique regional player for climate governance

Among the parties to the United Nations Framework Convention on Climate Change (UNFCCC), the European Union (EU) is the only supranational body, which provides another level of coordination and legislation. Indeed, the EU acts as a single party in international climate change negotiations, adding to the levels at which EU progress must be monitored: community and member state. Three main strategies are at the heart of the EU's decarbonisation policy: the 2020 climate and energy package; the 2030 climate and energy framework; and the 2050 low-carbon economy roadmap.

The 2020 climate and energy package is binding legislation to ensure that the Union meets its climate and energy targets for the year 2020. It sets out the

20-20-20 targets, i.e. a 20% cut in greenhouse gas emissions (from 1990 levels), a 20% share of EU energy from renewables, and a 20% improvement in energy efficiency.

The 2030 climate and energy framework also includes three key targets: a 40% cut in greenhouse gas emissions (from 1990 levels), a 27% share of EU energy from renewables, and a 27% improvement in energy efficiency.

Finally, the 2050 low-carbon economy roadmap represents the EU's long-term vision towards a 'virtually zero' emissions society. The roadmap is solely focused on cutting emissions, aiming at an aggregate reduction of 80% (from 1990 levels) and an intermediate target of 60% by 2040. The 2050 low-carbon economy currently represents a mere 'statement' and most of its framework is still under development.

In light of the Paris Agreement and the challenging actions it entails, the European Commission (EC) is presently updating its long-term decarbonisation strategies for 2050 and beyond, and is seeking public consultation in the revision process. The new strategy will need to consider how the EU can meet the ambitious 1.5°C target, which has not yet been reflected in its current climate policies.

EU targets provide an important indicative framework, but the Emissions Trading System (ETS), as a market instrument, and regulatory instruments, such as the Renewable Energy Directive, more concretely structure the Union's pathway towards a low-carbon future. Whereas the ETS applies one instrument across the EU, community regulations acknowledge the needs for tailored strategies in each member state.

The national, sub-national, and local pathways needed to achieve these international and national decarbonisation targets are much more complex than international agreements let on. Indeed, any such high-level policies are not meaningful if lower-level contexts are insufficiently understood. Top-down approaches, in terms of both policy choices and policy analysis, insufficiently reflect the realities on the ground. They should be complemented by bottom-up approaches that consider the unique circumstances of a varied group of stakeholders – particularly in order to understand barriers to the implementation of relevant policy choices, as well as their potential negative consequences and positive side effects. Fully understanding such risks and respective opportunities requires a diverse and flexible set of methods, and deep involvement of stakeholders. This is because a risk is not only context dependent but also largely depends on its perception, i.e. the eye of the beholder. Risk categories drawn from a single discipline therefore only constitute a limited subset of possible risks. These are often risks for which data is available and can therefore be quantified in terms of mathematical probabilities or variance. However, because they neglect risks that cannot be grasped by computable data, they may not be able to provide a complete picture for decision making or may insufficiently explain decisions made.

This book attempts to illustrate the diverse set of risks and uncertainties – both quantifiable and non-quantifiable – associated with low-carbon transition pathways, and thus contribute to informed decision making and improved risk management. This is particularly important considering the need for a transparent and sustainable decision-making process that leads to socially just and legitimate outcomes that are also politically and economically acceptable.

Following this introduction, Chapter 2 provides the conceptual background for our understanding of risks and uncertainties. This framing is innovative in its simplicity and its interdisciplinary, cross-thematic applicability. It is necessary to understand some of the terminology in the narrative chapters, but not an essential read to understand the main messages conveyed. Our framing of risk and uncertainty builds on the premise that risks are always context specific, perhaps more so than the transition pathways with which they are associated, and that risk perception plays a key role in understanding, communicating, and mitigating risks. It is therefore important that we use clear and complete descriptions of the risks we talk about.

In Chapters 3–13 we introduce transition pathways from countries across the globe, which derived from research processes integrating socio-economic modelling and stakeholder engagement efforts. The narratives presented here do not cover all possible contexts exhaustively, but rather attempt to present a broad range of pathways demonstrating the variety of actions needed at various levels within different cultural contexts, in order to deliver on the decarbonisation objectives of the Paris Agreement. Indeed, every country, every sub-national region even, will have specificities that make it unique. We thus attempt to achieve diversity in terms of political setting, socio-economic context, main levels of governance, and technological innovation systems.

Each narrative provides the relevant context to understand the current state of policy in the specific area discussed ('Where are we?'), the respective aims for the sector ('Where do we want to go?'), and the risks and uncertainties associated with the policy choices that could get us there ('How do we get there?'). The inter- and transdisciplinary methods used to obtain these narratives are kept short and distinct from the main narratives themselves, in order to create a more fluent reading experience. The respective detailed methodological descriptions can be found elsewhere for further consultation.

The 11 narratives are organised into four Parts that represent four main themes: (1) pathways for the transition of large-scale incumbent industry systems; (2) pathways towards renewable electricity systems; (3) pathways for energy efficiency in the building sector; and (4) pathways focusing on renewable energy technologies at household and community levels. The themes indicate the main level of analysis, but all consider (inter)action at multiple scales of governance, particularly with respect to risk perception of relevant stakeholders. As we will see in the next few paragraphs, some narratives fit more than one thematic description.

Under the first theme (Part II), we explore **pathways for the transition of incumbent large-scale technology systems** in Austria (Chapter 3), Canada (Chapter 4), and the United Kingdom (UK) (Chapter 5). Many countries across the globe still have dominant incumbent industries at the heart of their markets, which are strongly rooted in society and provide employment to large numbers of people. They can prove difficult to touch for politicians. Apart from the cases described here, this is, for example, the case with countries that invest heavily in fossil fuels, such as Poland (coal), Australia (coal), and Brazil (oil). Subjecting such industries to major changes with many associated uncertainties is economically and politically difficult. However, many of these industries are major contributors to climate change. They have also reached their limits with respect to incremental emission reductions, such as increasing efficiency. If they want to play a role in a low-carbon future, they face more fundamental changes, such as replacing coal as the major source of energy in the steel-making process.

In this book, we describe three large-scale industrial pathways and their associated risks and uncertainties. For Austria, we focus on energy supply for the iron and steel sector. For Canada, we discuss the Alberta oil sands, including the Athabasca oil sands from the perspective of the Indigenous people affected. For the UK, we examine the expansion of the nuclear power sector, which is a realistic option, vis-à-vis a nuclear phase out in favour of renewables. Kenya and Indonesia, featured in Part V, also illustrate regional and national technological lock-ins, complementary to local-level transitions. Both countries currently rely heavily on coal and oil for electricity production. In Indonesia one replacement for fossil fuels is biogas-for-electricity production, whereas Kenya has large geothermal resources as an opportunity to replace fossil energy.

The second theme (Part III) includes **pathways towards renewable electricity systems** in Chile (Chapter 6), the Netherlands (Chapter 7), Spain (Chapter 8), and Switzerland (Chapter 9). Solar power technologies have developed with incredible speed over the past two decades, with solar photovoltaics (PV) in particular becoming a competitive alternative to fossil fuels. As a standalone technology, or in combination with wind power, it has become an integral part of most national energy strategies. However, a range of barriers may still hinder successful implementation, and potential negative consequences resulting from implementation at scale are still insufficiently understood. In this book, we discuss the risks and uncertainties of transitioning to a solar-based electricity system for Chile, considering particularly its potential effect on energy poverty. The Netherlands' narrative includes the development of solar power and explores two alternative pathways, one centred on a large-scale solar parks, the other on small-scale rooftop installations. The Spanish case study offers lessons learned from the past: as a frontrunner in solar power technology and other renewable energy, the country was also among the first to experience unsuccessful strategies and their consequences. Finally, Switzerland illustrates the challenges in actually implementing a high-level strategy in a decentralised and multi-level democratic system, and potential to scale up renewable energy technologies in light of a planned renewable-based nuclear phase out.

In theme three (Part IV), we explore **energy efficiency pathways for the building sector**, illustrated with cases in China (Chapter 10) and Greece (Chapter 11). Improving energy efficiency can be most cost-effective and thus a principal measure to reduce emissions. However, it can involve rebound effects where increased efficiency leads to increased energy consumption, reducing its potential impact. In particular, the building sector is a popular target that holds immense potential. We explore China's very different path to more energy-efficient buildings. Here, the focus is naturally on the country's ever-growing cities, which provide an intriguing narrative on increasing energy efficiency at an unprecedented scale. We also look at one European case, Greece, which has a large remaining energy efficiency potential in its building stock but is struggling to achieve its near-term targets. The Greek narrative highlights how short- and long-term action requires different considerations for risk and uncertainty assessments.

Finally, theme four (Part V) covers **pathways focusing on renewable energy technologies at household and community levels** in Indonesia (Chapter 12) and Kenya (Chapter 13). At the end-user level, individuals and households constitute the most fundamental levels of decision making, and the perception of end-user changes at the small scale can often bely their importance. Behaviour at these scales can be the among the most challenging changes required to realise a low-carbon future.

For Indonesia, aside from large-scale biogas for electricity, we explore the potential to increase farm-level and communal biogas use in Bali by means of biogas digesters and cookstove technology. In Kenya we similarly draw attention to cookstove technologies, which need modernisation towards both more efficient technologies and alternative fuels in order to reduce deforestation and emissions.

The narrative style allows us to highlight the national, regional, and local specificities of each low-carbon transition pathway. The common focus on risks and uncertainties adds value in enabling us to draw conclusions across cases. The synthesis chapter (Chapter 14) highlights the lessons learned across all cases, emphasising the complexities in risk and uncertainty assessments with stakeholders, the ambiguity of any risk categorisation, and the joint challenges we face across the globe.

Overall, this book aims to contribute to the transdisciplinary dialogue on low-carbon transitions, particularly for energy. It offers narratives on some decarbonisation pathways in order to illustrate the associated risks and uncertainties. It also aims to encourage the uptake of this format of knowledge sharing, which ideally enables mutual understanding and learning and ultimately helps us make more salient, credible, legitimate, and – most importantly – sustainable decisions.

Note

1 Our narratives have the benefits of, but a broader basis than, narrative inquiry, which is a social–empirical (observational) research method exclusively based on an individual's experiences (Clandinin and Connelly, 2000); this and similar methods have been critiqued (Wood, 2000) for a lack of analytical definition and trustworthiness.

References

Cash, D.W., Clark, W.C., Alcock, F., Dickson, N.M., Eckley, N., Guston, D.H., Jäger, J., Mitchell, R.B. (2003). Knowledge systems for sustainable development. *Proceedings of the National Academy of Sciences* 100, 8086–8091. https://doi.org/10.1073/pnas.1231332100.

Clandinin, D.J., Connelly, F.M. (2000). *Narrative inquiry: experience and story in qualitative research.* Jossey-Bass, San Francisco, Calif.

Rounsevell, M.D.A., Metzger, M.J. (2010). Developing qualitative scenario storylines for environmental change assessment. *Climate Change* 1, 606–619.

UNFCCC (2015). *Report of the Conference of the Parties on its twenty-first session, held in Paris from 30 November to 13 December 2015. Addendum: Decisions adopted by the Conference of the Parties.*

UNFCCC (2016). *Report of the Conference of the Parties on its twenty-second session, held in Marrakech from 7 to 18 November 2016. Addendum Part Two: Action taken by the Conference of the Parties at its twenty-second session.*

Wood, D.R. (2000). Review: Narrative Inquiry: Experience and Story in Qualitative Research. *Anthropol. Educ. Q.* 31.

2 Framing risks and uncertainties associated with low-carbon pathways

Susanne Hanger-Kopp, Alexandros Nikas, and Jenny Lieu

This chapter introduces a broad conceptual framework for assessing risks and uncertainties across disciplines and subject matter. The framework enables transdisciplinary analysis for understanding risks and uncertainties associated with low-carbon energy transition pathways based on a consistent application in different country contexts. For this purpose, we require flexible epistemologies. We acknowledge that risks are socially constructed. Albeit related, we distinguish risk as a narrower concept from uncertainties. Within the social realm, we find some level of agreement on what important risks are, such as death, negative impacts on health, and excess loss of money. Some of these risks we can quantify, for example as probabilities or point estimates. Risks that cannot be quantified but which can be specified by stakeholders are no less important. Even concerns over issues that are low-risk according to quantitative probabilistic assessment need to be taken seriously, as they may have serious impacts on human behaviour, decision making, and acceptance.

Some economic approaches allow for risk to refer to positive impacts, i.e. opportunities (Markandya *et al.*, 2018). In this book we focus on the negative connotation of risk because we believe that this aspect may be neglected, given that low-carbon energy technologies, and associated policies and practices, are often considered to be inherently positive. However, in order to guarantee their legitimacy and lasting success, risks need to be anticipated and appropriately managed.

Risk and uncertainty

First, we introduce the baseline definitions for risk and uncertainty, which we find useful for inter- and transdisciplinary approaches, in order to identify and assess risks and uncertainties in the domain of decarbonisation policies. We believe that these may apply to other subject matter but this remains to be investigated. Second, we provide a short overview of relevant disciplinary approaches and state of the art discussions on uncertainty and risks, with the aim of identifying common ground to support interdisciplinary communication. We do not question disciplinary conventions on uses of the concepts, but rather suggest areas where leniency may be required to allow for inter- and trans-disciplinary understanding.

Risk and uncertainty are often used synonymously, as if they were one concept; this is particularly the case in the climate change research and policy-making community, and in the context of mitigation policy. For instance, in the most recent Intergovernmental Panel on Climate Change (IPCC) assessment report, working group III introduces risk management as a key pillar of the assessment report (IPCC, 2014a, 2014b), and the IPCC overall provides detailed guidance on how to address uncertainty in its work (Mastrandrea *et al.*, 2011). However, even in this co-ordinated assessment, we find several different interpretations of risks and uncertainties. This is potentially a spill-over from economics, which is among the most influential disciplines in this research area of climate–economy interactions and modelling, but where no one standardised understanding of either concept is provided.

At least for analytical clarity, but also for clarity in communicating across scientific boundaries, it is worth acknowledging and explaining how we avoid using these two concepts as synonyms in this book. Most often academic work, including pieces of the 'grey literature', when talking about 'risks and uncertainties' actually only address one of the two concepts. While this is not wrong, this interchanging use of the two terms may hinder effective academic work in support of climate policy. It may also cause confusion to both policymakers and those new to the study of risk and/or uncertainty.

We understand uncertainty as a general lack of knowledge of possible outcomes and states of the world. Uncertainty is a broader concept than risk, which refers to a calculated or perceived potential for negative impacts when the outcome is uncertain. Thus, uncertainty always retains an element of risk and, conversely, risk is always a form/subcategory of uncertainty. Risk is a reduced form of uncertainty, and one that we can better manage than most other forms. When we present disciplinary approaches to uncertainty, we only consider those that go beyond risk.

Perspectives on uncertainty

Since Socrates and then well into the seventeenth century, awareness of uncertainty and ignorance had been recognised as the key to wisdom. However, Descartes heralded three centuries of 'ignorance of ignorance', that is complete confidence in rationalism to provide certainty/truth, which became the normal state of the educated classes of Europe, particularly those in science (Ravetz, 2009, pp. xv–xvi). It was only in the second half of the twentieth century that a renewed awareness of uncertainty and ignorance returned. Thus, Ravetz argues, modern dealings with uncertainty have come to the fore with probabilistic risk assessment in the 1950s related to nuclear technology. Indeed, most concrete disciplinary work has been done in the context of risk as one specific form of uncertainty, as discussed in the following section. Broader ideas of uncertainty are rare and lack a disciplinary home. Some useful categorisations may help better distinguish and understand such uncertainties based on the extent of uncertainty, the sources of uncertainty, and the ability to measure uncertainty.

Former US Secretary of Defense Donald Rumsfeld famously distinguished that there are:

> known knowns; there are things we know we know. We also know there are known unknowns; that is to say we know there are some things we do not know. But there are also unknown unknowns – the ones we do not know that we do not know. And if one looks throughout the history of our country and other free countries, it is the latter category that tend to be the difficult ones.
>
> (Rumsfeld, 2012)

It is the realm of the known unknowns that is most accessible to researchers, and thus the subject of most academic discussion.

Within the realm of the known unknowns, uncertainty may result from different sources. Engineers, for example, broadly distinguish between uncertainty that can be reduced (epistemic uncertainty) and uncertainty that cannot be reduced as it is in the nature of things, such as the throwing of dice (aleatory uncertainty). Contributors to the latest assessment report of the IPCC (2014b), in addition to epistemic uncertainty as a result of a lack of information, consider paradigmatic uncertainty (resulting from disagreement about the framing of a problem) and translational uncertainty (resulting from incomplete or conflicting scientific findings).

Quantitative models can address uncertainty mathematically, most often by means of discrete scenarios that comprise specific values for the multiplicity of uncertain parameters, or other times by assigning joint distributions to uncertain parameters. The latter is usually done by means of sensitivity analysis – assessing effects of changing variables – and Monte Carlo simulations, which are essentially multiple random model runs. Qualitatively, uncertainty is more difficult to address. The IPCC, for example, uses qualitative rating scales for scientists to express the level of (un-)certainty which they associate with the key findings in their assessment report. First, scientists rate the validity of findings based on the type, amount, quality, and consistency of evidence. Second, they rate the findings probabilistically, based on statistical analysis of observations or model results, or expert judgement (Mastrandrea *et al.*, 2011). In the case of a negative outcome, we would talk about a risk assessment, according to the framing provided in this chapter. Both approaches assume that uncertainty is objective and external to a specific social reality.

By contrast, Smithson (2009) highlights the constructed nature of uncertainty and proposes a more flexible distinction based on how people talk about uncertainty (the nature of uncertainty), what they think it is (motives and values associated with uncertainty), and how they deal with it (coping/management). Such fully positivist and constructivist approaches are mutually exclusive. This remains true for risk.

Disciplinary approaches to risk

Most risk work has its origins in one of four subject areas: insurance and finance; natural hazards and disasters; technological innovation; or health. It has been inspired by multiple disciplines, such as mathematics, economics, psychology, and sociology, in an effort to reduce and manage uncertainty (Renn, 2008). Since the 1980s, we observe a shift from mathematical- and engineering-based work towards approaches stemming from social sciences, which is in line with the expansion of risk analysis from risk assessment to risk management and finally to risk communication. This represents a move towards the idea of subjective risk perceptions (Glickman and Gough, 1990).

Natural science approaches to risk are, for example, causal models in medicine and chemistry. Causal models test safety-related issues with organic and artificial substances by means of experiments. Mathematics provide the basis for actuarial and probabilistic risk assessment, which found its practical applications in the insurance sector and in the context of natural hazards. It is in these areas where natural sciences and social sciences, particularly economics, find common ground.

The natural sciences consider risks as objective and independent of social contexts. The social sciences assess risk either through the individual or through the collective. The economics of risk address risk taking and risk attitudes at the level of the individual. With their seminal work, and departing from the expected utility model, Kenneth Arrow and John Pratt laid the groundwork for an extensive literature on individual behaviour towards risk and risk aversion (Machina and Viscusi, 2014, p. xxxi). Indeed, the approaches of behavioural economics largely coincide with psychological methods of risk analysis, which also consider risk at the level of the individual.

The psychology of risk is in stark contrast to much of the mathematical risk analysis in the natural sciences and finance (Raue, Lermer, and Streicher, 2018). It traditionally followed a cognitive approach, assuming rational choosers who are constrained by their capacity to reason and learn (bounded rationality). Value-expectancy theory, the theory of reasoned action, and the theory of planned behaviour have been frequently used in psychological risk (Ajzen, 1991; Wigfield and Eccles, 2000). Classical methods such as experiments, observations, and interviews have been complemented with psychometric methods to measure risk perception and risk-related behaviour, which are also popular in behavioural economics (Slovic, 2000). The main tools here are standardised survey instruments. They are frequently criticised for systemic biases they may create, and sociologists have insisted on the importance of more in-depth methods to understand risks. Discourse analysis and in-depth interviews are the key methods used for this in sociology.

The sociology of risk, unlike economic and psychological approaches, applies at the collective level and is grounded in the assumption that risk is socially constructed. It purports that cultural biases and practices across social groups are effective drivers of ideas about risk and risk management. Prominent approaches are the Cultural Theory of Risk (CT) of Douglas and Wildavsky (1983) and Beck's Risk Society (1986).

Taylor-Gooby and Zinn (2006) highlight that economic, psychological, and sociological approaches are increasingly overlapping. They also emphasise the potential for cross-disciplinary fertilisation, particularly across psychological and sociological approaches but also with economics. This leaves bridging the gap between positivist and constructivist thinking. We find middle ground following a critical realist approach (Bhaskar, 2008; Sayer, 2000), where we create space for subjective risk perceptions and assessment, recognising the importance of concern (Fischhoff, Hope, and Watson, 1990) while conceding that within this social realm there are certain risks and assessments that can be considered objective, in the sense of a common understanding across stakeholder groups.

An interdisciplinary framework for assessing risks and uncertainties

The simplest common denominator to define risk across all these disciplinary approaches would describe it as the probability, chance, or potential for a negative outcome, impact, or consequence. Any further specifications of risk will always be context specific. Indeed, Fischhoff, Hope, and Watson claim that:

> the definition of risk, like that of any other key term in policy issues, is inherently controversial. The choice of definition can affect the outcome of policy debates, the allocation of resources among safety measures, and the distribution of political power in society.
>
> (Fischhoff, Hope, and Watson, 1990, p. 30)

In this book, we are interested in the risks associated with low-carbon transition pathways – or more precisely in the risks associated with the policy choices which constitute the foundation of such transition pathways – by promoting or mobilising the necessary transitions, principally with energy pathways. Frequently, such risks are termed as policy risks. Figure 2.1 illustrates how policy risk consists of two fundamental categories, which we call 'implementation risk' and 'consequential risk'.

- Implementation risk refers to the potential for diverse causes to affect the design, implementation, or success of a given policy; it introduces potential failure of policy implementation to the idea of barriers or challenges.
- Consequential risk refers to the potential for a certain policy to cause diverse negative consequences. Any anticipated consequential risk may become a cognitive barrier to the implementation of a pathway, i.e. an implementation risk.

Making this distinction explicit, and thus highlighting the importance of clearly describing 'cause' in addition to 'likelihood' and 'effect' for any given risk,

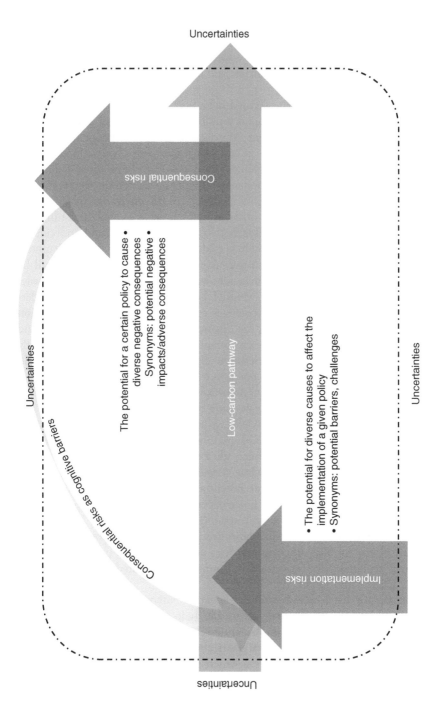

Uncertainties

Consequential risks

The potential for a certain policy to cause diverse negative consequences
Synonyms: potential negative impacts/adverse consequences

Low-carbon pathway

The potential for diverse causes to affect the implementation of a given policy
Synonyms: potential barriers, challenges

Implementation risks

Consequential risks as cognitive barriers

Uncertainties

Uncertainties

Uncertainties

Figure 2.1 Risks associated with climate mitigation policy.

constitutes the strength of this framing. Conventionally, such a distinction is only implied and remains opaque for anybody who is not familiar with the specific context.

The ideas of implementation and consequential risks easily translate to more conventional language, such as barriers and negative impacts, which is useful in dialogues with stakeholders.

For the sake of clarity, the stylised depiction of risks and uncertainties in Figure 2.1 omits several more complex aspects. First, the cause of any implementation risk, and the negative effect of any consequential risk, are not isolated but can be traced back and forth in cascades respectively. For example, public acceptance/opposition as an implementation risk may have several different root causes, such as experience, fundamental values, or common social practice, to name just a few. Similarly, negative impacts on the environment may result in negative impacts on health or other ecosystems services, which in turn may lead to economic implications for both economic sectors and individuals. Second, the framework does not address trade-offs and benefits. In any decision-making effort, benefits will play a role as well as risks, and trade-offs will be made. We neglect this in the framework as we assume that low-carbon pathways are mostly designed with their eventual benefits in mind, and that underestimating potential negative outcomes is one of the greatest implementation risks to any such pathway. Third, the distinction between implementation and consequential risks does not always completely hold. For instance, a badly implemented pathway may have negative impacts that affect the pathway itself, which has been the case in some emissions trading schemes.

Across all disciplines we find consensus that risk is context dependent. To account for this, the overarching framing for each narrative is based on the human innovation systems (HIS) idea, a framework developed based on the extended approach of the technology innovation system (TIS) framework (Hekkert et al., 2007). The framework explicitly acknowledges the spatial context and dimension of time, while considering the interaction between the natural ecological system (resource potential and limitations) and the human system. A low-emission pathway choice depends on the interaction between the environmental context (ecological system) and the HIS that encompasses social, economic, technological, administrative, and political contexts. Within the HIS there are also differing priorities at various scales (community, city, national, regional, etc.) governed by formal and informal institutions comprising actors with individual preferences.

Categorising barriers and potential negative outcomes in low-carbon pathways related to the contextual factors can help to conceptually disaggregate broadly defined categories of implementation and consequential risks, and specifically identify the problem. Systematically thinking through each category may help to identify the point from which the problems or barriers stem (e.g. lack of monetary funds, social resistance, level of technological readiness, etc.). The exercise can then help to identify the actors, institutions or policies that can address the barrier and prevent negative outcomes. These context categories are not clearly distinct from each other and inevitably overlap, for example in social and economic impacts. Thus, flexibility and at the same time clarity in the specific case-based application are paramount.

Methods for assessing risks and uncertainties

All narratives presented in this book draw on quantitative modelling, tightly interlinked with stakeholder consultations, to develop the transition pathways at their heart. The general equilibrium and optimisation models used for this purpose provide outputs as point estimates rather than probabilities, creating a false level of precision sometimes known as deterministic uncertainty (Nikas, Doukas, and Papandreou, 2018). Models suffering from this issue may be unsuitable if one wants to produce strategies that are less vulnerable to a large range of plausible outcomes. Moreover, for these models it is more difficult to consider implementation than consequential risk. It was thus of utmost importance to involve stakeholders not only for the specification of pathways but also for the identification and assessment of associated risks (Doukas *et al.*, 2018). For this purpose, we designed stakeholder engagement processes, all of which built on the same set of fundamental data collection formats, including: desktop research for the identification and analysis of policy documents; open and semi-structured expert- and stakeholder interviews (face-to-face, e-mail, telephone); and different focus groups and workshop formats. At the stakeholder identification and selection stage, we made sure that different, even opposing perspectives were considered in the analysis, so that different views could be heard and compared. This made the stakeholder consultation an as-good-as-possible representation of the actual fields of tension in our case study contexts. With stakeholders, we discussed risks and uncertainties qualitatively in interview and workshop settings, and with the option of a standardised assessment using Likert scales. This facilitated the characterisation of risks along different dimensions such as likelihood, severity of impact, timing of impact, mitigation capacity, and general level of concern.

Any of these social-empirical methods comes with certain limitations or potential biases, which can be reduced but not eliminated. For the best understanding and interpretation of our results, readers should be aware of the most important biases that potentially affect our research. These are:

(1) the selection of stakeholders: although the aim in each case was to select a comprehensive set of stakeholders using a stakeholder matrix approach, it is often not possible to include all relevant voices; and
(2) the capacity of stakeholders to assess risks, as stakeholders may or may not be experts with respect to the policies in questions.

Therefore, this is not a formal expert elicitation process (Morgan, 2014) aiming at quantified probabilities for very specific events, but an informed assessment of perceived potential barriers and negative outcomes. This is particularly the case for those risks and uncertainties that depend on individual and collective behaviour. As said previously, during stakeholder elicitation it is important to identify concerns and areas of disagreement across stakeholder groups in order to arrive at a broad qualitative assessment of transition pathways. Expert elicitation could then be useful as a potential follow-up on individual risks, where this is possible. An important aspect to keep in mind in this respect is that the consultations were not aimed at generating objective information from stakeholders, as there is no such thing as an objective point of view, particularly with respect to risks and uncertainties. The main goal, instead, was to make these points of view explicit so that they can be considered and valued in the development of the transition pathways.

The complementarity of the modelling and stakeholder approaches allowed for a broad consideration and detailed understanding of risks, taking into account the different perspectives of individuals and/or stakeholder groups. While the details of the modelling aspects of these pathways and the particular interactions with stakeholders are published elsewhere (Dalla Longa and van der Zwaan, 2017; e.g. Lieu *et al.*, forthcoming; van der Zwaan *et al.*, 2018). The narratives in this book provide a non-technical and in-depth approach to the respective low-carbon pathways. These are based on the findings from the extensive stakeholder engagement processes used in the respective case studies. Many of these use one or more specific stakeholder-driven methods for identifying risks, uncertainties, opportunities, and benefits in the specific national contexts. These include system mapping (SM), fuzzy cognitive mapping (FCM), and multiple-criteria group decision making (MCGDM).

SM (Albu and Griffith, 2006) is a qualitative mapping method recently applied to the climate change and policy domain (Nikas *et al.*, 2017). It constitutes a novel approach to defining system boundaries, capturing interactions among institutions and technological value chains and describing enabling environments for these value chains, such as cultural aspects, existing policies and regulation, and habits. Based on the mapping, system inefficiencies or other obstacles can be identified that potentially block the successful implementation of the transition pathway: implementation risks. Moreover, the mapping can reveal potential opportunities within a system to benefit from when pursuing the transition. The identified risks and opportunities may directly or indirectly facilitate or hinder a transition pathway, or a policy aiming to promote that pathway. Once elicited from policymakers and other relevant actors, this knowledge is translated into a diagrammatic form. In this book, the SM framework is based on the components of a TIS: enablers and barriers affecting the development of a system include policy instruments, contextual factors, institutions, and infrastructure. The method is widely used in the following chapters to help inform integrated assessment modelling and other policy support activities.

With a more quantitative approach, FCM is a semi- or quasi-quantitative modelling technique that essentially comprises a visual (mapping) component and a mathematical (simulation) component. It aims to identify how a particular change to the system is perceived to affect it. This change can be a socio-economic driver or a policy instrument, the impacts of which are useful to assess when seeking the optimal policy strategy. As a semi-quantitative modelling framework, FCM cannot offer insights into the exact outcomes resulting from a specific change to a system. The assessment is, however, meaningful in a comparative manner, offering insights into how a policy instrument or mix is perceived by stakeholders compared to all other policy instruments or mixes.

In a similar fashion to system mapping, the visual component of FCM involves designing a map of concepts into which different systems processes can be broken down. However, this map is intended to represent only relationships that express influence, causality, and system dynamics. Based on the captured causality, the system is then simulated in order to rank the alternative policy instruments or mixes in terms of positive influence on the system variables representing the ultimate objectives (Nikas and Doukas, 2016).

Finally, MCGDM is a sub-discipline of operational research, aimed at supporting decision making in complex problems where multiple views (decision makers) on multiple dimensions (criteria) must be considered before reaching a solution. Multi-criteria analyses have long been used to support decision making in energy (Doukas, 2013) and climate policy (Nikas, Doukas, and Martínez López, 2018). In the context of this book, MCDGM is based on the TOPSIS method (Hwang and Yoon, 1981) and is used for assessment of implementation and consequential policy-related risks against a consistent family of evaluation criteria. The criteria include: the likelihood to manifest; the level of impact on the policy framework (for implementation risks) or the severity of impact (for consequential risks); and the mitigation capacity, as perceived by stakeholders.

Final remarks

The fundamental outline and categorisation above provide us with an interdisciplinary language for identifying and assessing risks and uncertainties which can easily be translated to practical, inter- and transdisciplinary needs in the narratives presented in this book. The meaning remains the same, while the terminology and other context-specific elements can be adapted. Its value lies in its simplicity and transparency. Transparency, as we make clear where we must deviate from disciplinary approaches at either end of the wide spectrum of risk assessment methods, most importantly mathematics and sociology. The simplicity lies in our approach to describing risk, where we emphasise and realise the need to describe clearly the fundamental components of each risk identified. Ultimately, this approach enables us to synthesise and compare our findings with respect to the risks and uncertainties associated with each pathway.

References

Ajzen, I. (1991). The theory of planned behavior. *Organ. Behav. Hum. Decis. Process.*, *Theories of Cognitive Self-Regulation* 50, 179–211. https://doi.org/10.1016/0749-5978(91)90020-T.

Albu, M., Griffith, A. (2006). Mapping the market: participatory market-chain development in practice. *Small Enterp. Dev.* 17, 12–22. https://doi.org/10.3362/0957-1329.2006.016.

Beck, U. (1986). *Risikogesellschaft. Auf dem Weg in eine andere Moderne*, 22. ed. Suhrkamp, Frankfurt am Main.

Bhaskar, P.R. (2008). *A Realist Theory of Science.* Verso, London; New York.

Dalla Longa, F., van der Zwaan, B. (2017). Do Kenya's climate change mitigation ambitions necessitate large-scale renewable energy deployment and dedicated low-carbon energy policy? *Renew. Energy* 113, 1559–1568. https://doi.org/10.1016/j.renene.2017.06.026.

Douglas, M., Wildavsky, A. (1983). *Risk and Culture: An Essay on the Selection of Techno-logical and Environmental Dangers.* University of California Press.

Doukas, H. (2013). Modelling of linguistic variables in multicriteria energy policy support. *Eur. J. Oper. Res.* 227, 227–238. https://doi.org/10.1016/j.ejor.2012.11.026.

Doukas, H., Nikas, A., Gonzalez-Eguino, M., Arto, I., Anger-Kraavi, A. (2018). From Integrated to Integrative: Delivering on the Paris Agreement. *Sustainability* 7.

Fischhoff, B., Hope, C., Watson, T. (1990). Defining risk, in: *Readings in Risk. Resources for the Future*, Washington D.C., pp. 30–42.

Glickman, T.S., Gough, M. (1990). *Readings in Risk.* Resources for the Future.

Hekkert, M.P., Suurs, R.A.A., Negro, S.O., Kuhlmann, S., Smits, R.E.H.M. (2007). Func-tions of innovation systems: A new approach for analysing technological change. *Technol. Forecast. Soc. Change* 74, 413–432. https://doi.org/10.1016/j.techfore.2006.03.002.

Hwang, C.-L., Yoon, K. (1981). *Multiple Attribute Decision Making: Methods and Applica-tions – A State-of-the-Art Survey. Lecture Notes in Economics and Mathematical Systems.* Springer-Verlag, Berlin Heidelberg.

IPCC (2014a). *Climate Change 2014: Impacts, Adaptation, and Vulnerability. Part A: Global and Sectoral Aspects. Contribution of Working Group II to the Fifth Assessment Report of the Intergovernmental Panel on Climate Change*, Field, C.B., V.R. Barros, D.J. Dokken, K.J. Mach, M.D. Mastrandrea, T.E. Bilir, M. Chatterjee, K.L. Ebi, Y.O. Estrada, R.C. Genova, B. Girma, E.S. Kissel, A.N. Levy, S. MacCracken, P.R. Mastrandrea, and L.L. White (eds.). Cambridge, United Kingdom, and New York, NY, USA.

IPCC (2014b). *Climate Change 2014: Mitigation of Climate Change. Contribution of Working Group III to the Fifth Assessment Report of the Intergovernmental Panel on Climate Change*, Edenhofer, O., R. Pichs-Madruga, Y. Sokona, E. Farahani, S. Kadner, K. Seyboth, A. Adler, I. Baum, S. Brunner, P. Eickemeier, B. Kriemann, J. Savolainen, S. Schlömer, C. von Stechow, T. Zwickel and J.C. Minx (eds.). Cambridge University Press, Cambridge, United Kingdom, and New York, NY, USA.

Lieu, J., Hanger-Kopp, S., Sorman, A., van Vliet, O. (forthcoming). Special Issue: Assessing risks and uncertainties of low-carbon transition pathways. *Environ. Innov. Soc. Transit.*

Machina, M., Viscusi, K. (Eds.) (2014). Introduction, in: *Handbook of the Economics of Risk and Uncertainty.* North-Holland, pp. xxxi–xxxv. https://doi.org/10.1016/B978-0-444-53685-3.00022-2.

Markandya, A., Sampedro, J., Smith, S.J., Dingenen, R.V., Pizarro-Irizar, C., Arto, I., González-Eguino, M. (2018). Health co-benefits from air pollution and mitigation costs of the Paris Agreement: a modelling study. *Lancet Planet. Health* 2, e126–e133. https://doi.org/10.1016/S2542-5196(18)30029-9.

Mastrandrea, M.D., Mach, K.J., Plattner, G.-K., Edenhofer, O., Stocker, T.F., Field, C.B., Ebi, K.L., Matschoss, P.R. (2011). The IPCC AR5 guidance note on consistent treat-ment of uncertainties: a common approach across the working groups. *Clim. Change* 108, 675. https://doi.org/10.1007/s10584-011-0178-6.

Morgan, M.G. (2014). Use (and abuse) of expert elicitation in support of decision making for public policy. *PNAS* 111, 7176–7184. https://doi.org/10.1073/pnas.1319946111.

Nikas, A., Doukas, H. (2016). Developing Robust Climate Policies: A Fuzzy Cognitive Map Approach, in: Doumpos, M., Zopounidis, C., Grigoroudis, E. (Eds.), *Robustness Analysis in Decision Aiding, Optimization, and Analytics*. Springer International Publishing, Cham, pp. 239–263. https://doi.org/10.1007/978-3-319-33121-8_11.

Nikas, A., Doukas, H., Lieu, J., Alvares-Tinoco, R., van der Gaast, W. (2017). Managing stakeholder knowledge for the evaluation of innovation systems in the face of climate change. *J. Knowl. Manag.* 21.

Nikas, A., Doukas, H., Martínez López, L. (2018). A group decision making tool for assessing climate policy risks against multiple criteria. *Heliyon* 4, e00588. https://doi.org/10.1016/j.heliyon.2018.e00588.

Nikas, A., Doukas, H., Papandreou, A. (2018). A Detailed Overview and Consistent Classification of Climate-Economy Models, in: *Understanding Risks and Uncertainties in Energy and Climate Policy: Multidisciplinary Methods and Tools towards a Low Carbon Society*. Springer, Berlin.

Raue, M., Lermer, E., Streicher, B. (Eds.) (2018). *Psychological Perspectives on Risk and Risk Analysis: Theory, Models, and Applications*. Springer International Publishing.

Ravetz, J. (2009). Preface, in: Bammer, G., Smithson, M. (Eds.), *Uncertainty and Risk: Multidisciplinary Perspectives*. Earthscan.

Renn, O. (2008). Concepts of risk: An interdisciplinary review – Part 1: Disciplinary risk concepts. *GAIA* 17, 50–66.

Rumsfeld, D.H. (2012). U.S. Department of Defense News Briefing – Secretary Rumsfeld and Gen. Myers.

Sayer, R.A. (2000). *Realism and social science*. Sage, London; Thousand Oaks, Calif.

Slovic, P. (2000). *The Perception of Risk*. Earthscan, London; Sterling, VA.

Smithson, M. (2009). The Many Faces and Masks of Uncertainty, in: Bammer, G., Smithson, M. (Eds.), *Uncertainty and Risk: Multidisciplinary Perspectives*. Earthscan.

Taylor-Gooby, P., Zinn, J.O. (2006). Current directions in risk research: New developments in psychology and sociology. *Risk Anal.* 26, 397–411. https://doi.org/10.1111/j.1539-6924.2006.00746.x.

Wigfield, A., Eccles, J.S. (2000). Expectancy–Value Theory of Achievement Motivation. *Contemp. Educ. Psychol.* 25, 68–81. https://doi.org/10.1006/ceps.1999.1015.

van der Zwaan, B., Kober, T., Longa, F.D., van der Laan, A., Jan Kramer, G. (2018). An integrated assessment of pathways for low-carbon development in Africa. *Energy Policy* 117, 387–395. https://doi.org/10.1016/j.enpol.2018.03.017.

Part II

Pathways for incumbent large-scale technology systems

3 Austria

Co-designing a low-carbon transition pathway focusing on energy supply for the iron and steel sector

Brigitte Wolkinger, Jakob Mayer, Andreas Tuerk, Gabriel Bachner, and Karl Steininger

The EU's decarbonisation goals for 2050, as laid down in the *Roadmap for moving to a competitive low-carbon economy in 2050*, aim for reducing greenhouse gas (GHG) emissions by 80 to 95% (compared with 1990 levels) and requires steep emission reductions after 2030 (EUR-Lex–52011DC0112–EN–EUR-Lex, n.d.). So far, decarbonisation efforts have relied on gradual improvements in energy efficiency or expansion of renewables, but over coming decades there is a need to accelerate efforts through breakout technologies and cost reduction. De Pee *et al.* (2018) emphasise in their analysis that decarbonisation pathways for the industrial sector are less defined (than, for example, when compared with the power sector) as the sector is embedded in a global context and has a relatively high share of emissions from feedstocks and high-temperature heat. In Austria, GHG emissions from energy production, transport and other energy sources in total decreased between 1990 and 2014 by about 5% while emissions from industry (including process emissions) increased by 12% (European Environment Agency, 2016). Besides energy efficiency measures, further and stronger mitigation and more radical action is required, such as substituting current industrial processes with low-emission alternatives. For some industrial sectors this could be the most fundamental change of the production process in their history.

Industrial production, and especially basic material production such as iron and steel, is often strongly interrelated with other sectors of the domestic and international economy. Its centrality in economic production has been empirically tested and detected in several regional economies deploying network theory on input–output tables (e.g. for the US, Canada, and Mexico by Aroche-Reyes (2002) and by Muñiz, Raya, and Carvajal (2008) for the case of Europe). This implies a relatively wide range of possible risks and indirect effects when decarbonising these key sectors. However, as technical options are being developed, the associated risks of their implementation are often insufficiently assessed because the scope of analysis for such options is often set very narrowly (e.g. Fischedick *et al.*, 2014) and thus salient aspects like system-wide effects and repercussions are disregarded. A quantitative assessment of the macroeconomic effects of iron and steel decarbonisation is presented in Mayer, Bachner, and Steininger (2019). This chapter aims to make an important qualitative contribution to this discussion by revealing risks perceived and stated by different stakeholder groups, which might emerge when it comes to the implementation of 'climate neutral' steel production technologies, taking into account the interrelated transition of the energy supply sector.

In Austria, the iron and steel industry and energy supply sectors comprise nearly half of the country's GHG emissions, both within the ETS (Emissions Trading System) and outside the ETS. These two sectors contribute 16% to real gross domestic product (GDP). Within Austria's economy, iron and steel form 15.5% of total GHG emissions and contribute 2% to domestic value added (Anderl *et al.*, 2017a; Statistics Austria, 2018b). The iron and steel sector also stands out because of the high risks associated with a low-carbon transition due to its exposure to geo-political developments, its dependence on global (feed-stock) markets (especially for coal and gas, iron ores and scrap), the extraordinarily long economic lifetime of investments, and prevailing overcapacity issues. For the steel industry there is little scope for significantly reducing carbon emissions while continuing with the current production method (Mayer, Bachner, and Steininger, 2019). The sector therefore faces several options: either its replacement with new business activities based on alternative materials such as wood or polymers (demand-side measures); a switch to hydrogen-based steel production (as fuel, feedstock or in electrolysis); the use of zero-carbon electricity; the use of biomass for feedstock (charcoal) and carbon capture and storage (CCS); or carbon capture and use (CCU) – or a combination of all these options. Hydrogen-based steel production is a major consideration for Austria's largest steel producer. As this would require a large amount of zero-carbon electricity, the transition of the Austrian electricity sector is being systematically considered as an interrelated pathway.

The role of the iron and steel sector in a low-carbon transition

Austria's long tradition of high-quality steel production renders the case of a low-carbon transition for the sector particularly complex. Since 1970 steel output has doubled, now amounting to eight million metric tons per year. Each ton of steel produced is associated with *process* emissions which emerge from chemical reactions during processing of iron ores ('oxygen reduction'), thus they differ from *combustion* emissions. The theoretical minimum process emission intensity of the currently applied technology, a blast furnace (BF), is about 1.3 tons of CO_2 per ton of steel (Scholz *et al.*, 2004; Kirschen, Badr, and Pfeifer, 2011), with Austrian production being slightly above that at 1.5 tons of CO_2 per ton of steel (Anderl *et al.*, 2017b). Hence, further efficiency efforts would be restricted to a maximum reduction of 13% in CO_2 process emissions. Although the emission intensity of a single ton has significantly improved over recent decades, total CO_2 process emissions have not declined since output of the steel sector increased much faster. Hence, the sector is still a major contributor to the national GHG inventory (Anderl *et al.*, 2017a). With the theoretical lower bound of emission intensity being insufficient for reaching long-term climate objectives, it is apparent that more radical mitigation options have to be explored, such as: (i) a major steel output decline requiring substitutes with similar versatile product characteristics for specific applications, e.g. wood (composites) or polymers; (ii) the deployment of best-available technology options combined with carbon capture and storage/usage; (iii) turning to secondary steel production with sufficient high-quality scrap feedstocks; or (iv) fuel switches from carbon-intensive to no-carbon processes with competitive relative unit costs for comparable steel grades. The most relevant side constraint for each of these four options is indicated here in a simplified way. (For a more detailed discussion see Mayer, Bachner, and Steininger, 2019). Recently, Austrian steel producers, in particular Voestalpine AG (the largest steel-producing company in Austria and one of the leaders worldwide), have acknowledged the last option to be one possible pathway to follow, which would require switching to new processing technologies within the coming two decades (i.e. hydrogen-based instead of coke-based oxygen reduction of iron ores), thereby rendering deep decarbonisation of the sector possible (Mayer, Bachner, and Steininger, 2019).

The necessary provision of sufficient renewable electricity for hydrogen production puts pressure on the electricity sector. However, it is possible (and intended) to store hydrogen (either on site or by a third party) and to use it on demand in iron and steel production. For Austria, and according to stakeholders, electrolysis-based hydrogen generation would mean 33 TWh (terawatt hours) per annum of electricity devoted to the iron and steel industry in order to maintain the national steel output at current levels. Current total electricity generation in Austria amounts to about 60 TWh, having doubled since 1970. Hence the electrification of this single industry is perceived by stakeholders to necessitate a massive expansion of the domestic electricity generation system. In Mayer, Bachner, and Steininger (2019) and Bachner *et al.* (2018a; 2018b) we emphasise that this 'additionality' in terms of increased domestic electricity generation can be questioned because the implementation of hydrogen-based iron and steel technologies triggers relative price and thus foreign trade effects which lead to lower 'additional' domestic electricity supply and demand than anticipated bottom-up. Likewise, mitigation efforts are endangered if electricity demand is provided carbon intensively, which would merely lead to a shift of emissions from the steel industry to the energy supply sector.

Investigating the current electricity mix, hydropower clearly dominates electricity production in Austria, comprising about two-thirds of generated electricity, with fossil-fuel-based generation representing about 18% (in Austria about 75% of its electricity production is from renewables (Oesterreichs Energie, 2018)). In this context it is worth noting the Federal Constitutional Act for a Nonnuclear Austria (Federal Constitutional Act for a Nonnuclear Austria, 1999), which prohibits the construction and operation of installations for the production of energy by means of nuclear fission. Moreover, in recent years

Austria lost its status as a net exporter of electricity. This contextual framework raises questions around self-sufficiency, further options for the expansion of low-carbon generation (e.g. cost-competitive spots for hydropower complying with environmental legislation are almost completely exploited), the coupling of historically separated sectors, as well as other electrification trends (mobility, digitalisation, etc.).

Moreover, industry representatives have frequently confronted climate policy measures with concerns about potential added value losses due to decreasing competitiveness against competitors in non-policy regions, caveats regarding carbon leakage, as well as unfavourable job market implications. However, evidence on the consequences of unilateral environmental regulation like carbon pricing is mixed. Particularly, the dynamics such unilateral measures may set in motion are diverse as, for instance, suggested by the hypothesis of Porter and Linde (1995). A recent review by Dechezleprêtre and Sato (2017) finds statistically significant adverse effects of environmental regulation which, however, are limited to the short run and are comparably small because production, trade, and choices of location are shown to be mostly determined by other factors such as sunk capital and costs of transportation. However, this review mainly builds upon observations from the past, where carbon prices were moderately low, thus evidence against a plausible threshold level of carbon pricing for which adverse effects of unilateral environmental measures dominate is lacking. Consequently, re-location measures of heavy industries remain a prevalent threat in public controversies.

The point of departure for a low-carbon transition

This chapter gives a brief overview of current socio-economic, technological, and political framework conditions that have to be considered when moving towards a low-carbon society. The insights are based on a stakeholder dialogue described in the *Research process and methods* box.

Research process and methods

In Austria a range of projects and stakeholders address future low-carbon energy scenarios. The broad transition pathways covered in this chapter, and an understanding of the related risks, were developed applying different methods of stakeholder participation.

First, semi-structured interviews were conducted to gain information on the Austrian socio-economic and political context related to a low-carbon transition.

In two stakeholder workshops visions for a low-carbon society in 2050 and associated transition pathways were developed, with corresponding risks and uncertainties assessed and prioritised. In the *first workshop* the specific Austrian context was discussed and transition pathways towards a low-carbon society in 2050/2030 were developed using the back-casting approach, i.e. starting from a desired future. Risks and uncertainties were developed along these pathways in an interactive group workshop setting. Two main dimensions to be considered in the pathways emerged, which were: i) a detailed description of the needed technological change and a timeline for implementation of mitigation technologies for the two focus sectors, as well as their future sectoral interrelationships; and ii) the institutional framework and corresponding milestones, emphasising the institutional and political needs for initiating and staying on the decarbonisation track until 2050.

Throughout the process of development of transition pathways, about 100 risks and uncertainties were worked out by stakeholders and the research team in an iterative co-creation process (*first workshop* and *bilateral calls*). These risks were first clustered for further processing and ranking in the second stakeholder workshop, in a World Café setting, and comprise implementation as well as consequential risks. Although stakeholders were asked about these different kinds of risk in the first round, going into detailed exploration of the core problem and the question behind each risk (as well as ranking the risks) is much easier by thinking in thematic complexes.

Bilateral calls between the two workshops were also used to refine and discuss assumptions and results of the specific pathways (for the steel and iron and corresponding electricity pathway), which were entered into the macroeconomic model for assessing impacts.

The feedback on *macroeconomic assessments* were essential for deploying the WEGDYN computable general equilibrium (CGE) model (Mayer, Bachner, and Steininger, 2019) and the E3ME econometric model (Barker *et al.*, 2012) to inform and quantify selected consequential risks.

Furthermore, a *survey* was conducted to evaluate risks by different criteria (e.g. impact of risk, timing, and probability).

The most important participating stakeholder groups were from industry (the iron and steel, cement, petrochemical, and innovative technologies for renewable energy), power supply companies, the chamber of labour (as part of the 'social partnership' which is part of the Austrian consensus-based political system), ministries (Environment, Finance), political parties (only from the Green Party; other parties were invited but did not attend the workshop due to elections), and NGOs. Our approach has been to collect risks over a very broad spectrum by inviting relevant stakeholders from diverse fields. Hence, we only focus on different (categories of) risks and rarely differentiate between perceptions. While the risks for industry were not sufficiently clear to other stakeholders, societal risks, such as distributional effects, were highlighted by NGOs or the chamber of labour. Instead of necessarily finding a consensus on all major risks, the range of risk was illustrated to provide a basis for policy design, as Austrian policymakers so far have a limited understanding of the risks related to a low-carbon transition.

Austria is a country with a strong industrial base and a large electricity generating sector. The latter has a high share of renewables, although there is still potential to increase this. The country is highly developed and wealthy, with large natural resources and a highly educated population. In principle, it has the institutional and financial prerequisites needed for a transition to a low-carbon economy and to be a driver and frontrunner within the EU.

Awareness of a need for change is presently strong among policymakers. However, most political parties fear to discuss and promote climate change issues, arguing that there would be a threat of deindustrialisation (this has been repeatedly discussed in the context of the European Emissions Trading Scheme between major industrial players in Austria and policymakers). Short planning horizons driven by budgetary constraints are also a limiting factor here. These factors result in stagnation at the policy level and lack of awareness regarding the degree of needed reforms. The country's new government came up with an updated climate strategy in early 2018 (Ministry of Sustainability and Tourism and Ministry of Transport, Innovation and Technology, 2018); however, this lacks a concrete timetable for transition. Climate and sustainable development are not yet sufficiently recognised as public issues affecting employment, equality, or education, and a corresponding public discourse is missing. Despite these uncertainties with respect to public action, it can be observed that sections of Austrian industry take more action than actually required by policymakers. In a project called 'H2Future', the leading Austrian steel company, the leading power company, and international partners are developing cross-sectoral solutions for a hydrogen-based production (Voestalpine AG, 2017). In a further H2020 project called Steelanol, the Austrian service provider Primetals Technologies supports a European steel maker in commercially operating one of the first industrial-scale plant using greenhouse gas emissions from steel production as feedstock for bio-ethanol production (Steelanol, 2018).

There are possible diverging interests between national and provincial governments due to Austria's federalist administration, as well as opposing views in Austria's 'social partnership'. This social partnership is an influential block of constitutional interest groups that are involved in policymaking that is able to block mitigation measures. Spatial planning policy, for example, which is a local and regional competence, has caused irreversible and inefficient structures that Austria has to deal with. These structures cannot be ignored when designing low-carbon transitions but have to be integrated into a future governance system. Introducing new technologies, such as electric vehicles, only makes sense when reforms in spatial planning have been introduced first in order to avoid redundant transportation. However, political resistance on the local and regional level for changing current legislation is significant and counteracts national decarbonisation strategies. Thus, continuing policies that favour urban sprawl not only increase transportation but also imply massive infrastructural requirements when large electric vehicle penetration rates are targeted, e.g. by implementing a charging network.

Elements of the transition, such as for a more decentralised energy supply in Austria, are slowly emerging and gaining acceptance. However, technological and socio-economic shifts are also necessary. According to stakeholders' perceptions, technologies that are able to initiate transitions are already available but are not implemented on a large scale due to uncertainties, which are described in detail later on. Currently, alternative lifestyle scenarios with absolute reductions in demand for products and services, and thus energy throughput, are not being discussed in Austria. A societal problem that may hinder the transition is the increasing income gap between cities and rural areas, as well as between rich and poor (real income in the lower segments has declined over the last 20 years). The lowest quartile of equivalent income spends 5.7% on energy consumption while the highest quartile only spends 3.9% (Statistics Austria, 2018a). Possible increases of energy prices will thus be borne to a larger extent by the lower-income group. Stakeholders also question the access to new technologies for low-income groups.

Process emission-free iron and steel transition pathway and corresponding low-carbon electricity pathway

The desired low-carbon future and the corresponding transition pathways to get there, as developed by stakeholders, are here synthesised and described for the iron and steel sector and the energy sector.

The desired low-carbon future as a starting point for transition is characterised by a highly efficient, yet still energy intensive, complex industry that produces specialised, complex products. Companies are almost free of waste and there are closed material cycles (including carbon). Most of the energy used is renewable electricity, both domestic and imported (e.g. wind electricity from northern Europe or PV from northern Africa). There are centralised and decentralised elements (storage as well as production) which are fully integrated and controlled via digital technologies and artificial intelligence. Mobility is fully electrified and multimodal. There are no individual cars anymore; instead cars are shared and co-ordinated via artificial intelligence. People live in denser spatial patterns, especially in cities, with short distance ways and in comfortable plus-energy houses. Steel is likely to be still produced in Austria and in the same quantities as today, but with a hydrogen-based technology with hydrogen being produced via renewable electricity (power to gas). Research co-operation on hydrogen production and storage between the main Austrian steel producer, the largest energy supply company, and the regulator ensures Austrian electricity supply are successful. Policymakers on a national level are supporting these efforts. Steel production with gas (methane) is planned as a bridging technology.

Technological advancements, however, will support not only decarbonisation but also improvements in general the quality of life. Austria will be characterised by a high R&D quota and innovation. In the industrial sector, centralised solutions will still be needed as there is need for large amounts of energy. Decentralised solutions are linked to stronger citizen participation regarding a low-carbon economy. Therefore, national, regional, and local policymakers will play a crucial role in ensuring these prerequisites for companies, individuals, or collectives producing renewable electricity.

In order to reach the desired future of deep decarbonisation in the steel and iron sector, different technological options are proposed: hydrogen-based direct reduced iron (DRI) and plasma-direct-steel-production (PDSP). Assuming different timing of implementation and electricity costs, the macroeconomic model revealed impacts on GDP and employment (Mayer, Bachner, and Steininger, 2019).

A promising transition pathway not only needs industry (e.g. steel production) to be electrified, but also a broader electrification of a range of other economic activities including, for example, e-mobility. The transition towards an electrified economy is thus simultaneously a precondition for the transition of the steel sector. There is consensus among stakeholders in Austria that energy supply in 2050 will be 100% renewable (PV, wind, H_2 produced from renewables) and that new opportunities from digitalisation will lead to a highly energy-efficient economy. Part of the transition pathway in order to reach this target are assumptions on the sectoral electricity generation portfolio under specific socio-economic developments. The socio-economic development included in modelling is based on the shared socio-economic pathway framework (SSP2) (O'Neill *et al.*, 2014). For the 100% renewable target, additional investments for storage (power to gas and batteries) are becoming necessary from 2036 onwards. This has been modelled by a macroeconomic model which in turn is based on a proposed pathway by Pleßmann and Blechinger (2017) with refinements by stakeholders (especially for the wind capacity).

The broader vision

The draft transition pathway for the iron and steel sector was also seen as a broader pathway to move to a hydrogen economy by 2050. It considers bridging solutions, such as natural gas, that can cause lock-in effects, as well as upstream value chain requirements (e.g. renewables-based hydrogen generation).

As the need for hydrogen in sectors other than steel is still unclear, possible supply chains are not well understood, posing an important risk. Stakeholders highlighted that inter-sectoral relationships should, however, be actively explored as part of a circular economy. Demand for hydrogen produced via renewables for process emission-free steel production might compete with hydrogen needs in other sectors of the economy (e.g. the chemical industry, mobility). Relevant caveats apply with regards to the simultaneous transition in the electricity supply subsystem towards renewables in order to prevent a shift from process emissions in the iron and steel sector towards combustion-based emissions in power generation.

Austrian stakeholders suggested rethinking the concept of 'decarbonisation', rather aiming for 'carbon management', as carbon itself might still play a crucial role in a future economy. Carbon management is distinct from a fully decarbonised future as the focus is on the management of the balance of GHG emission sources and sinks. It allows for CO_2 reduction by using intelligent grids and electricity from renewables in combination with chemical storage to avoid the fluctuations of renewables. Carbon-intensive production and consumption is in principle permitted; however: (i) the atmospheric net balance should not be positive, and (ii) this intent can be facilitated by achieving an as-far-as-possible decarbonised economic system. Besides sequestering, remaining carbon emissions can be methanised with green hydrogen, stored and traded in gas tanks/pipeline systems. In this circular economy, material and energy cycles are well adjusted to each other and mostly closed.

New processes can convert CO_2 into CH_4 which can be re-used as basic material. Also, materials such as concrete can play a role in storing energy. Stakeholders emphasised the crucial importance of pursuing (in parallel) the exhaustion of energy efficiency potentials, the change of the energy mix, and the implementation of storage technologies.

Austria is embedded in the European Union which has developed towards more integration and the capacity to act commonly, particularly facilitating cross-border action within the EU (e.g. harmonisation measures intended for an 'Energy Union'), as well as negotiating more effectively with larger regional blocks such as China or the USA. There is already significant EU action to stimulate a low-carbon transition until 2020 and to provide a reliable regulatory framework. At the EU level, a strategy will be developed that enables electricity production and transfer across countries. Given the large quantities of electricity that Austria would need in the midterm if steel production is based on hydrogen, stakeholders stress that domestic potentials are insufficient. Also, import possibilities might decline since neighbouring countries might also choose electrification pathways. South-eastern Europe with its large potential regarding renewables could – under consideration of nature protection requirements – provide potential for imports to Austria. Besides the possibility of importing electricity, Mayer, Bachner, and Steininger (2019) and Bachner *et al.* (2018a; 2018b) show that deep decarbonisation of Austrian steel making is within domestic renewable potentials.

As an essential aspect of a transition towards decarbonisation of economic activities, or of carbon management, in Austria by 2050, a deliberate institutional framework is needed, plus an end to any public support for fossil fuels.

In 2050, Austria is net climate neutral and resilient overall to ecological, social, and economic challenges. Co-operation between the private and the public sector is strongly developed. 'Know how' exchange and co-operation in the triangle of public services, the economy, and sciences are made via a dedicated climate secretariat and supported by a centre for climate studies. In the industrial sector there will be refineries, in particular for supplying non-energetic by-products. CCS is allowed in the meantime as part of a carbon management (as described above). Policy measures are therefore designed in an integrative way and a fragmentation of different interests is reduced.

Risks and uncertainties related to the pathways

In the stakeholder co-production process, clusters of risks have been created in order to get a more holistic picture of risks both before and during implementation of the Austrian transition pathways (Figure 3.1). These clusters and accompanying sub-clusters comprise implementation risks (barriers) as well as consequential risks (impacts). While some clusters intrinsically focus

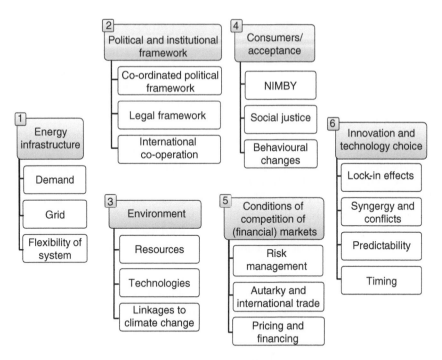

Figure 3.1 Risk clusters for the transition in the iron and steel sector and the energy sector.

on implementation risks only (like cluster 2), others contain mainly consequential risks (like cluster 3).

Cluster 1: Energy infrastructure

Many debates came up among different stakeholder groups when *energy infrastructure* clusters of risks and uncertainties were discussed. Changing demand for
electricity from renewables, challenges for the grid (stability), and the issue of
flexibility of the energy system are involved in this cluster. A controversial issue
was which segments of society will carry the burden of a transition in the energy
sector. A well-designed energy infrastructure, be it centralised or decentralised,
is a prerequisite for the energy transition and therefore involves many implementation risks in this cluster. However, risks can also occur after or during the
transition, like the supply and price volatility of renewable energy.

Misleading market rules and market design (pricing and regulations) were
mentioned as the main barriers for transition in the energy sector. Market over-
regulations and subsidies on fossil fuels were criticised as well, and so was the
current electricity market not including the grid.

For the energy-intensive iron and steel industry, the question of sufficient
electricity supply is key for the transition, with likely increasing electricity

demand from other sectors potentially interfering with security of supply. When switching to the hydrogen-based direct reduction route (with subsequent processing of pig iron in electric arc furnaces), large amounts of electricity are needed. The way in which the future energy supply (and demand) system is designed determines the scale and composition of auxiliary infrastructure, like transmission and distribution grids, or gas pipelines and storage caverns. For instance, it is not currently clear which party will generate and eventually supply hydrogen to the market – it may eventually be the currently incumbent power utilities, the iron and steel sector itself, or a providing third party. It is also conceivable that the generation of electricity and/or hydrogen is organised in a much more decentralised manner, such that on-site generation of 'green hydrogen', for example, decreases the upstream dependence of the iron and steel sector.

Along the pathways there is the risk that the sudden increase in electricity demand from renewable energy in Austria or Europe cannot be met by potentials, or that the supply side has not been prepared to meet it. This risk could increase substantially if large industries decide to switch to carbon-neutral electrified processes and/or individual transport switches to e-mobility. A second branch of consequential risks concerns the electricity grid with different perceptions. While some stakeholders expect an increase in grid instability due to the volatility of renewable energy (especially when expansions in subordinate grids are lagging), others perceive the risk as being very low due to intelligent digital solutions. In addition, there is a risk that the system is not flexible enough to cope with expected variations in energy supply from renewables and that storage capacities are too small. This risk increases when a 100% renewables-based electricity mix is targeted and should include insights from increased efficiency of co-generation.

There is no consensus among stakeholders concerning the system design of future electricity infrastructure. Some favoured an expansion of central power grids to transfer large amounts of power away from renewable power plants, especially in peak times. Others argued that such investments are sunk costs, and that one should start with decentralised energy from a very local level in response to the electricity demand there, then supply the electricity demand of the next level. This would also increase the local or regional added value (e.g. engineers, plumbers, etc.). However, a central–decentral mix is seen as reasonable depending on the overall energy supply system design (cf. the cases with on- or off-site 'green' hydrogen generation for the iron and steel sector).

Cluster 2: Political and institutional framework

Another cluster of mainly implementation risks hindering, postponing, or jeopardising transition is endemic in the *political and institutional framework*. Here we saw a broad consensus. While on a meta-level there is the risk of lacking or missing co-ordinated national initiatives, on a very specific lower level legal regulations may be counterproductive. Furthermore, national transition depends on future international co-operation and developments, which are uncertain and bear risks.

First of all, there is no integrated climate and energy strategy in Austria with clear milestones and measures. An additional barrier for transition is the lack of a roadmap for hydrogen, which would be instrumental for solving the need for storage and reliable supply. This leads to uncertainty due to lack of liability, planning security, commitment on a clear target, and the timing of implementation of technologies. In Austria, responsibilities for one specific topic are spread over different levels of authorities (local, regional, and national) and therefore co-ordinated legislation is often blocked by differing interests.

Furthermore, the national transition depends on future international co-operation and developments. One important indicator for planning further climate mitigation projects is the development of a CO_2 price, which is currently uncertain, especially for governments of single EU member states since the design of the EU ETS (Emissions Trading Scheme) is ruled at the EU level. When the regulatory framework for a transition is missing or even counterproductive, a transition will not be initiated; for the private sector in particular, planning security is a prerequisite for financing investments. For instance, regulations for the use of decentralised electricity are unclear, as is stability for reinvestment in renewable energy sources (RES) and legal regulations within the clean power law. Various stakeholders stress that phase-out laws for fossil-fuel-based energy technologies are missing, and long approval procedures and time lags for expansion projects (e.g. hydropower), due to concerns around nature protection, are seen as barriers for fast transition by the energy sector.

Lack of international co-operation can bear the risk of an industry resettling towards regions with a less restrictive climate policy. A switch to new technologies in the European iron and steel industry will be influenced by either increased tariffs or cheap gas imports, making it less competitive to switch from conventional blast furnace routes to (currently more expensive) hydrogen-based routes. However, for the leading steel company in Austria, two-thirds of its revenue from operations in the USA (€1.2 billion in 2017) is generated as a local manufacturer in the USA itself (Voestalpine AG, 2018). Clear regulations and financial incentives from European energy policy would be necessary to switch to new technologies in the iron and steel sector, thereby increasing the value for the products.

Cluster 3: Environment

The *environment* cluster mainly sees occurrence of consequential risks. Due to the volatility of renewable energy, storage is increasingly important. This leads to increasing demand for resources and, inevitably, environmental consequences during extraction or disposal (e.g. for lithium which is used in batteries for storage of electricity). The magnitude and importance of this risk was disputed between energy producing utilities and NGOs. When transition pathways include CCS technology to be able to reach the climate targets, respective consequential risks are borne. For example, induced seismic activity may lead to leakage and thus to (in)direct human/animal suffocation (through soil acidification). In addition to expected intermittency, there might be negative feedback loops with climate change affecting the renewable energy supply, e.g. for hydrogen power when rivers dry up. Increasing resistance to the construction of new overhead high-voltage power lines could be a barrier to transition projects or impose increased costs for power supply companies along the transition pathway via the need to install underground cables. The disposal of infrastructure, e.g. of wind turbine rotors which are made of composite material which are not easily disposed of, can be seen as further consequential risks along the transition pathway which are often not considered because only the energetic flows enter into the analysis.

Cluster 4: Consumer/acceptance

The *consumer/acceptance* cluster includes mainly implementation risks. The 'not in my backyard' (NIMBY) effect (e.g. van der Horst, 2007) refers to the risk of residents opposing renewable energy projects (e.g. hydropower due to nature protection concerns) is seen as a main implementation risk within this cluster. The neglect of social equity issues and the role of behavioural change along the transition pathways are further – mainly consequential – risks that should be considered.

Implementation risks opposing transition are the play-off between social justice and climate mitigation, lacking planning or investment security, and the role of behavioural change in transition. In Austria, PV and wind are gaining importance. There is not much remaining potential to construct large hydro power plants, while small power plant projects can receive opposition from nature conservationists. Hydropower will, however, definitely play a key role in domestic energy transition (e.g. pumped hydro). Stakeholders consider the risk that, in the public debate, social justice and climate mitigation are seen as contradicting targets. The fact that climate mitigation strategies can make an important contribution to social justice is not communicated sufficiently. For households, risks around investing in or planning their energy supply can lead to reluctance and uncertainty. For example, whether to invest in a private PV installation or a co-generation system depends on the legal framework and clear political commitment. In this context, there is the risk that private households make too high or unnecessary investments in storage technologies at home due to high feed-in tariffs. For instance, a 10kWh battery for at-home storing costs about €10,000, while an electric vehicle can store 40kWh while simultaneously being used for private mobility.

If all actors are not considered, there is the risk of energy poverty for those who have to pay potentially higher prices or who have no access to new technologies. One example of when this may occur is through net metering (a usage and payment scheme in which a customer generating their own power is compensated monetarily) if the additional cost of providing these schemes is borne by other electricity customers.

Another risk occurs before or during implementation of the transition pathway, when there is no incentive for behavioural change and its contribution is not sufficiently considered or communicated. Especially when transition pathways are mainly focused on technological changes, the contribution of behavioural change is not incorporated into projections. To overcome this risk, stakeholders suggest clearly communicating which parts of society are able and willing to contribute to behavioural change and the extent of change required. When changing behaviour (such as saving energy or producing electricity) is associated with status, then all parts of the society could be involved. For example, smartphones can be seen as a status symbol for all parts of society. To, for example, show one's power consumption and production in real time could become a trend, and be an incentive for private investment and behavioural change.

Cluster 5: Conditions of competition of (financial) markets

The cluster *conditions of competition of (financial) markets* includes mainly financial risks and the lack of planning security for investors as a prevalent implementation risk. Increasing dependence on electricity imports and international policies (e.g. steel policy in the USA) as well as the underestimation of market dynamics are seen as further risks within this cluster.

Lack of foresight and efficiency as the main criteria for decision and planning, as well as missing education for professions needed in future, are further risks that may hinder transition. If politics fosters transition, there is the risk that the intended direction will harm some sectors or companies in unintended ways.

The risks stemming from financial markets is appraised by stakeholders ten times higher than from the real economy. As decarbonisation is highly dependent on financial capital, developments in financial markets are very important for the transition. Additional risks occur for the proposed transition pathway in the iron and steel sector. For its implementation, electricity prices between €0.03 and €0.05/kWh, together with modest CO_2 pricing, guarantee a cost-efficient transition as they level out unit operating expenditures of hydrogen-based steel production with conventional carbon-intensive production (details are given in Mayer, Bachner, and Steininger, 2019). However, these prices are uncertain at the current state. Which sectors will compete for hydrogen produced by renewable electricity (the demand by mobility, housing, the chemical industry, etc.) is also unclear. The potential for a hydrogen economy exists, but emerging new dependencies and asymmetric market power relations may have to be dealt with. For instance, pipeline owners possess to some extent monopoly positions because hydrogen can be transported mainly by pipelines.

In general, corporate risk management always has to consider aspects like market concentration, foreign trade, predictability, and political interventions. More specifically, in the future risk management will change when the iron and steel pathway includes profound switches from coal, coke, or gas to electricity or hydrogen. This is because steel producers will interact in different value chains and on different markets with different risk profiles.

An additional consequential risk mentioned relates to increasing market penetration of electric vehicles. This could discriminate against the domestic/European automotive industry, which is currently producing cars with conventional combustion engines. Moreover, the net employment effect of such switches is unclear. For the given case, traditional jobs like mechanics might be put under pressure while electricians and IT technicians might gain.

Cluster 6: Innovation and technology choice

Risks in the cluster *innovation and technology choice* comprise consequential and implementation risks, e.g. the risk of lock-in effects for new technologies and corresponding path dependency, lack of information infrastructure (smart solutions), the risk of large investments for private households when investing in decentralised solutions, and the question of timing to finance an investment.

Another implementation risk refers to non-existing co-ordination across sectors for their technology choices. Furthermore, there is a lack of public financial support for innovation and research, which is especially critical for projects that are between pilot technology and industrial application (upscaling).

Path dependency and lock-in effects (e.g. through capacity mechanisms) are consequential risks along the transition pathway. A diversification of risk can be achieved by initiating different technological options. The risk of timing is one along the transition pathway but also a kind of implementation risk. For instance, the right time for installation of hydrogen-based technology to be competitive is very difficult to predict.

Prioritisation of risks by cluster

The comprehensive number of risks collected through the stakeholder process has been prioritised in a World Café setting. The scoring procedure has been done by risk cluster, i.e. the prioritised risks represent the most important risks for each cluster but cannot be compared across risk categories. They were then allocated to the implementation and consequential risks category. High-scored implementation risks are lack of a national strategy and co-ordinated European policy and the lack of a regulatory framework (thus poor timing of investments). On a more individual level, the play-off between climate mitigation and social justice and the role of behavioural change were also highly scored. High-scored consequential risks from a stakeholder perspective are the stability of grids and flexibility of the energy system, the possibility of lock-ins, and the fear that households bear the main part of the costs due to misleading signals from politics or markets. The risk that resources are not considered in transition calculations is seen as the most relevant risk for the environment cluster. (See Figure 3.2).

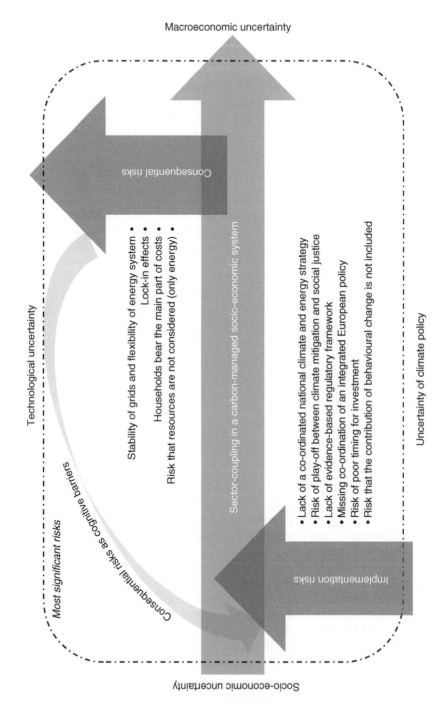

Macroeconomic uncertainty

Technological uncertainty

Uncertainty of climate policy

Socio-economic uncertainty

Consequential risks

Implementation risks

Stability of grids and flexibility of energy system
Lock-in effects
Households bear the main part of costs
Risk that resources are not considered (only energy)

Sector-coupling in a carbon-managed socio-economic system

- Lack of a co-ordinated national climate and energy strategy
- Risk of play-off between climate mitigation and social justice
- Lack of evidence-based regulatory framework
- Missing co-ordination of an integrated European policy
- Risk of poor timing for investment
- Risk that the contribution of behavioural change is not included

Most significant risks

Consequential risks as cognitive barriers

Figure 3.2 Main implementation and consequential risks associated with a transition of the Austrian iron and steel sector.

Besides stakeholder perceptions of the most relevant implementation and consequential risks, appraising and modelling impacts of the transition pathways and correlated risks can only be done in the light of uncertainties. Within the modelling process of transition pathways, several uncertainty layers were identified (Bachner *et al.*, 2018b). There are several technological options for transition and corresponding costs statements that are uncertain and lead to different effects along transitions. The possible developments of population and economic growth lead to socio-economic uncertainty and are based on the shared socio-economic pathways (SSPs) (O'Neill *et al.*, 2014). The stringency and geographic scope of climate policy is uncertain and highly influences other factors to be considered in modelling transition pathways. The uncertain macroeconomic state of a country along the transition pathway influences the degree of capacity utilisation.

Conclusions

The co-developed narrative presented in this chapter highlights the risks associated with disruptive technological changes that are implied by transition to low-carbon energy production for Austria's iron and steel sector. It illustrates the strong sectoral interrelationships and resulting risks, making the transition in the iron and steel sector dependent on the transition in the energy sector as well as other sectors. The technology choice in one sector will influence the transition pathway of the other sectors and vice versa; for example, whether mobility will be electrified or supplied by hydrogen will be crucial for the iron and steel sector.

Stakeholders from all kinds of affiliation stress that missing long-term policy frameworks, and the lack of cross-sectoral alignment of technology choices and supply chains, can hinder transition.

Technological innovation in Austria is currently strongly driven by the industry itself with public financial support (mainly by the EU), while the socio-political framework on the national level is not to date pushing for a substantial transition. This implies that industry would need to take risks with possible negative consequences but also with the possibility of becoming technological pioneers.

Another main implementation risk stressed by stakeholders is the trade-off between climate mitigation and social justice. If there is no clear policy framework that considers and mitigates threats posed to social justice by the transition, there may be fear and opposition towards changes. In addition, the question of who pays for the transition is fundamental here. The companies involved need planning security for investment and a clear regulatory framework. This is crucial for private investment in decentralised solutions. If it is not clear to 'prosumers' to what extent the investment is part of a higher-level energy strategy, they will restrain their investments. In our understanding, such long-term subsidy guarantees are part of comprehensive energy strategies. If information regarding such instruments is lacking, so will be planning security for prosumers. Yet the Austrian climate policy rather follows past patterns of gradual improvement and does not provide a clear framework that would reduce uncertainty and most of the implementation risks. Neither the opportunities nor the need for fundamental changes to the stable, wealthy, and consensus-seeking Austrian society are well enough understood.

While short-term goals or decisions are associated with risks, long-term goals are seen as introducing a broad set of socio-economic opportunities. A 'mission-oriented' approach could be a solution, i.e. a truly 'entrepreneurial state' which not only corrects market failures but also actively creates and shapes (new) markets through investing and signalling credibility: a state that consciously takes on risks that the private sector either will not or cannot accept. This, however, could lead to the 'public risk–private benefit' issue (Mazzucato, 2016). In particular, in the energy sector the low-carbon transition is seen as a way towards a better quality of life, a more inclusive way where citizens are more involved in the energy system beyond their narrow role in reducing emissions.

The stronger need for co-operation between the public and the private sector to mitigate transition risks will increase the overall societal resilience to ecological, social, and economic challenges. Sector coupling may also increase the efficiency of the economy and lead to overall carbon management of the economy instead of implementing individual and sectoral measures. Carbon management is more than just decarbonisation: it includes synchronisation of material and energy cycles in a sustainable circular economy, and allows for CO_2 emission reduction by using intelligent grids and electricity from renewables in combination with chemical storage.

With a new set of risks emerging, policies need to be designed in an inter-sectoral way. These policies will reduce and achieve more effective management of the risks in individual sectors but will also need new institutional formats and policy processes to design them. If Austria aims to remain or become a frontrunner in innovation, a detailed planning process for a risk-reducing climate and industrial policy far beyond the 2030 horizon needs to be initiated.

References

Anderl, M., Burgstaller, J., Gössl, M., Haider, S., Heller, C., Ibesich, N., Kuschel, V., Lampert, C., Neier, H., Pazdernik, K., Poupa, S., Purzner, M., Rigler, E., Schieder, W., Schneider, J., Schodl, B., Stix, S., Storch, A., Stranner, G., Vogel, J., Wiesenberger, H., Winter, R., Zechmeister, A. (2017a). *Klimschutzbericht 2017* [online]. Umweltbundesamt, Wien. Available at: www.umweltbundesamt.at/fileadmin/site/publikationen/REP0622.pdf [accessed 22 January 2018].

Anderl, M., Haider, S., Pazdernik, K., Schmidt, G., Schodl, B., Schwarzl, B., Titz, M., Weiss, P., Zechmeister, A., Stranner, G., Elisabeth Schwaiger, Pfaff, G., Pinterits, M., Poupa, S., Purzner, M., Schmid, C., Moosmann, L., Lampert, C., Kriech, M., Friedrich, A. (2017b). *Austria's National Inventory Report 2017. Submission under the United Nations Framework Convention on Climate Change and under the Kyoto Protocol* [online]. Environment Agency, Vienna, Austria. Available at: www.umweltbundesamt.at/fileadmin/site/publikationen/REP0608.pdf.

Aroche-Reyes, F. (2002). Structural Transformations and Important Coefficients in the North American Economies. *Economic Systems Research* 14(3), 257–273.

Bachner, G., Mayer, J., Steininger, K.W., Anger-Kraavi, A., Smith, A. (2018a). *The economy-wide effects of deep decarbonization and their uncertainties – The case of the European iron and steel industry*. EUH2020 project TRANSrisk working paper.

Bachner, G., Wolkinger, B., Mayer, J., Tuerk, A., Steininger, K.W. (2018b). *Risk Assessment of Low Carbon Transition Pathways for Austria's Steel and Electricity Sectors – Insights from a Co-Production Process*. EUH2020 project TRANSrisk working paper.

Barker, T., Anger, A., Chewpreecha, U., Pollitt, H. (2012). A new economics approach to modelling policies to achieve global 2020 targets for climate stabilisation. *International Review of Applied Economics* 26(2), 205–221. https://doi.org/10.1080/0269217 1.2011.631901.

Dechezleprêtre, A., Sato, M. (2017). The Impacts of Environmental Regulations on Competitiveness. *Review of Environmental Economics and Policy* 11(2), 183–206.

EUR-Lex–52011DC0112–EN–EUR-Lex (n.d.). Available at: https://eur-lex.europa.eu/legal-content/EN/TXT/?uri=CELEX:52011DC0112 [accessed 30 August 2018].

European Environment Agency (2016). *National emissions reported to the UNFCCC and to the EU Greenhouse Gas Monitoring Mechanism*. Available at: www.eea.europa.eu/data-and-maps/data/national-emissions-reported-to-the-unfccc-and-to-the-eu-greenhouse-gas-monitoring-mechanism-11 [accessed 21 September 2016].

Federal Constitutional Act for a Nonnuclear Austria (1999). *Federal Constitutional Act for a Nonnuclear Austria* 1. BGBl. I Nr. 149/1999 [online]. Available at: www.ris.bka.gv.at/Dokumente/Erv/ERV_1999_1_149/ERV_1999_1_149.html.

Fischedick, M., Marzinkowski, J., Winzer, P., Weigel, M. (2014). Techno-economic evaluation of innovative steel production technologies. *Journal of Cleaner Production* 84, 563–580.

van der Horst, D. (2007). NIMBY or not? Exploring the relevance of location and the politics of voiced opinions in renewable energy siting controversies. *Energy Policy* 35(5), 2705–2714.

Kirschen, M., Badr, K., Pfeifer, H. (2011). Influence of direct reduced iron on the energy balance of the electric arc furnace in steel industry. *Energy* 36(10), 6146–6155.

Mayer, J., Bachner, G., Steininger, K.W. (2019). Macroeconomic implications of switching to process-emission-free iron and steel production in Europe. *Journal of Cleaner Production* 210, 1517–1533. https://doi.org/10.1016/j.jclepro.2018.11.118.

Mazzucato, M. (2016). From market fixing to market-creating: a new framework for innovation policy. *Industry and Innovation* 23(2), 140–156.

Ministry of Sustainability and Tourism and Ministry of Transport, Innovation and Technology, 2018. *#mission 2030 – Die Klima- und Energiestrategie der Bundesregierung*. Available at: www.mission2030.bmnt.gv.at [accessed 12 April 2018].

Muñiz, A.S.G., Raya, A.M., Carvajal, C.R. (2008). Key Sectors: A New Proposal from Network Theory. *Regional Studies* 42(7), 1013–1030.

Oesterreichs Energie (2018). *Daten und Fakten zur Stromerzeugung*. Available at: https://oesterreichsenergie.at/daten-fakten-zur-stromerzeugung.html.

O'Neill, B.C., Kriegler, E., Riahi, K., Ebi, K.L., Hallegatte, S., Carter, T.R., Mathur, R., Vuuren, D.P. van (2014). A new scenario framework for climate change research: the concept of shared socioeconomic pathways. *Climatic Change* 122(3), 387–400.

de Pee, A., Pinner, D., Roelofsen, O., Somers, K., Speelman, E., Witteveen, M. (2018). *Decarbonization of industrial sectors: the next frontier* [online]. Amsterdam, Netherlands. Available at: www.mckinsey.com/~/media/McKinsey/Business%20Functions/Sustainability%20and%20Resource%20Productivity/Our%20Insights/How%20industry%20 can%20move%20toward%20a%20low%20carbon%20future/Decarbonization-of-industrial-sectors-The-next-frontier.ashx.

Pleßmann, G., Blechinger, P. (2017). How to meet EU GHG emission reduction targets? A model based decarbonization pathway for Europe's electricity supply system until 2050. *Energy Strategy Reviews* 15, 19–32.

Porter, M.E., Linde, C. van der (1995). Toward a New Conception of the Environment-Competitiveness Relationship. *Journal of Economic Perspectives* 9(4), 97–118.

Scholz, R., Pluschkell, W., Spitzer, K.-H., Steffen, R. (2004). Steigerung der Stoff- und Energieeffizienz sowie Minderung von CO_2-Emissionen in der Stahlindustrie. *Chemie Ingenieur Technik* 76(9), 1318–1318.

Statistics Austria (2018a). *Household Budget Surveys 2014/15*. Available at: www.stat.at/web_en/statistics/PeopleSociety/social_statistics/consumption_expenditures/household_budget_survey_2014_2015/index.html.

Statistics Austria (2018b). *Structural Business Statistics 2015 – Main Results*. Available at: www.statistik.at/web_en/statistics/Economy/industry_and_construction/structural_business_statistics/049989.html.

Steelanol (2018). *Home* [online]. Available at: www.steelanol.eu/en [accessed 30 August 2018].

Voestalpine AG. (2017). *Voestalpine, Siemens and VERBUND are building a pilot facility for green hydrogen at the Linz location – voestalpine* [online]. Available at: www.voestalpine.com/group/en/media/press-releases/2017-02-07-voestalpine-siemens-and-verbund-are-building-a-pilot-facility-for-green-hydrogen-at-the-linz-location/[accessed 27 August 2018].

Voestalpine AG (2018). *Trump's decision regarding Section 232 – a max. of 3% of voestalpine Group revenue in the USA impacted*. Available at: www.voestalpine.com/group/en/media/press-releases/2018-03-09-president-trumps-decision-regarding-section-23-a-maximum-of-3-percent-of-voestalpine-group-revenue-in-the-usa-impacted/ [accessed 12 April 2018].

4 Canada

Finding common ground – the need for plural voices in lower-carbon futures of the Alberta oil sands

Luis D. Virla, Jenny Lieu, and Cecilia Fitzpatrick

Sustainability in the Alberta oil sands

The knowledge of oil sands has been part of the Frist Nation cultural heritage for centuries, not as an energy source but as an isolation material for canoes. The first historical documents date to the eighteenth century when European explorers and crown emissaries, brought to the region by fur trading economic interests, observed and described the abundance of bitumen near the Athabasca River as a natural substance emanated from the ground (Canadian Association of Petroleum Producers (CAPP), 2016). One hundred years later, the first government-sponsored geological study on the Alberta oil sands was performed and, along with additional expeditions, unveiled a big economic potential for the use of the resource. During this time, natural gas and conventional oil were discovered, developed, and used as primary energy source for the communities in the province (Alberta Ministry of Culture and Tourism, 2016). However, commercial development of the Alberta oil sands did not begin until the early 1920s with some failed attempts to extract the oil using drilling wells between 1906 and 1917 (Regional Aquatics Monitoring Program, 2016). From this moment, technologies for oil separation from the sand using hot water were developed using the same principle as the First Nations people had used in the past. This technology was mainly applied at a pilot scale and was supported by governmental institutions such as the federal Department of Mines and the Alberta Research Council.

The main focus of Alberta oil sands exploitation by that time was roofing and road surfacing applications (Alberta Ministry of Energy, 2016). It was not until 1962 that the large-scale oil sands development began. That year, the government of Alberta developed a specific oil sands policy to supplement the conventional crude oil policy already in place. The inaugural project was the Great Canadian Oil Sands (GCOS) Project, later transformed into the current Suncor Energy, which brought the first oil sands operation 'on stream' in 1967. Along with the GCOS, the Syncrude consortium was formed and shipped its first barrel of oil in 1978, becoming the second major oil sands producer in Canada (Insitute for Oil Sands Innovation, 2016). Today, the oil sands attract local and foreign investments. The oil sands region in Alberta is divided in three major deposits – Athabasca, Cold Lake, and Peace River oil sands – which are the world's third-largest proven oil reserves (see Figure 4.1) (Alberta Energy Regulator, 2015). Two methods are vastly used to recover the oil sands deposits: surface mining and in-situ recovery. Most of the efforts are put on developing technologies to decrease production cost and the environmental impact of the oil sands while keeping a competitive price for exports to the international markets.

The Alberta oil sands in Canada are one of the biggest fossil-fuel energy resources in the world. For Canada, the fossil-fuel sector is both a major economic driver and source of greenhouse gas (GHG) emissions. By 2016, energy contributed to 9.9% of Canada's GDP, of which 2.2% was from crude oil (Natural Resources Canada, 2018). Canada is ranked as one of the world's top ten emitters, contributing 1.6% of global emissions (International Energy Agency, 2015). Fossil-fuel production and transportation are the largest contributing sectors to the total GHG emissions in the country, with 27% and 23% of contributions respectively (Natural Resources Canada, 2016). Alberta contributed 37.4% of national emissions in 2014, representing the biggest emitter among all Canadian provinces (Environment and Climate Change Canada, 2017).

Canada has the second largest oil reserves in the world, most of which primarily exist in the form of crude bitumen. The majority of proven reserves are found in oil sands in western Canada. Unproven reserves, however, are expected to be

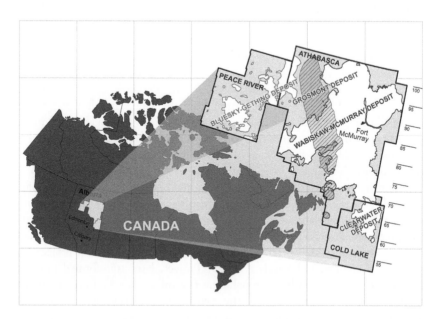

Figure 4.1 Location of Alberta's oil sands areas and selected deposits.

Source: Alberta Energy Regulator, 2015.

significantly larger and are likely to reside in Alberta's oil sands. Extracting crude bitumen, or synthetic crude oil, from the oil sands generates significant GHGs and other polluting emissions (methane, sulphur, aromatics) that has led to damaging environmental and health impacts. Historically, developing the oil sands and protecting the environment have been a challenge for policymakers and industry representatives (Chastko, 2000). After World War II, the oil and gas sector became the dominant economic industry in the province, and by the 1970s this industry accounted for 40% of the province's income. The rapid growth of the sector in the twenty-first century has also influenced politics since actions towards regulation of the industry could be perceived as detrimental to the economic progress of the province (Finkel, 2000). Today, as part of the Paris Agreement, Canada has agreed to decrease GHG emissions to 30% below 2005 levels by 2030, mainly by applying regulatory measures to change the transportation and electricity sectors, control landfill emissions, and promote clean energy technologies (Government of Canada, 2017). To ensure a successful implementation of different lower-carbon policies, it is critical for policymakers to account for socio-economic effects and social acceptance. Particular attention is needed to understand the needs and interests of the communities most vulnerable to the impacts of the Alberta oil sands production.

Oil sands companies claim that sustainability is an issue of importance in the sector. The industry has been making efforts to address the problem by developing new technologies to cut GHGs. From the technical perspective, sustainability is approached by air pollution monitoring, decreasing water usage, optimising existing processes, and land reclamation. The main oil sand producers have teams dedicated to sustainability issues. Presently, interactions with neighbouring communities and development programmes are included as part of their sustainability strategies. Although technical concepts of sustainability are being applied in the sector, many environmental degradation issues continue to persist. Decreased flow in the Athabasca River, increasing air pollution (claimed to be associated with increasing incidences of disease among the local population), shrinking caribou and moose ranges, and the disappearance of certain animal species are all a result of decades of resource development in the northern Alberta boreal forest (Tenenbaum, 2009; McLachlan, 2014; Parajulee and Wania, 2014; Russell, Pendlebury, and Ronson, 2016; Harner *et al.*, 2018).

The perspectives and interests of various sub-national groups most impacted by climate action and policies are not currently reflected in international and regional climate policies, such as the UN's 1.5–2.0°C target, Canada's Nationally Determined Contributions (NDC), and the Alberta Climate Leadership Plan. The exclusion of different perspectives leads to a roadmap that is disconnected from local realities and can threaten the implementation of climate change actions. The Paris Agreement specifically called to respect, promote, and consider the rights of local communities including Indigenous peoples among others. Indeed, pathways developed from views from Indigenous or local communities, in which generations of knowledge are embedded, can help inform policymakers on necessary actions to address climate change. The local narratives can be the basis of developing lower-carbon transition pathways that represent and address the needs and priorities of the Indigenous communities and complement other pathways defined by policymakers, industry players, and the public at large.

This chapter presents three lower-carbon transition pathways which are based on the priorities of policymakers, the community of Fort McKay First Nation, and a diverse mix of industry, grid operators, and NGOs in Alberta. Risks associated with these three pathways were elicited considering the perspectives of various types of stakeholders. These risks could act as barriers for the materialisation of the lower-carbon pathways, or as negative impacts from the implementation of them. In addition, the analysis of these risks sets the grounds for the development of effective policies in the midst of governmental changes in a Canada looking to meet its commitments in the Paris Agreement.

Reseach process and methods

The approach selected for the development of the narrative consisted of two stages: pathway development, and elicitation of preference and risks associated with the pathways. The broader public sentiment with regards to climate action policies and lower-carbon futures for the Alberta oil sands was collected through media articles. These included newspapers, blogs, and government, industry and institutional (NGOs, companies) websites. We also collected information from Statistics Canada (Canada's national statistical agency). The three pathways represent three distinct views originated from policymakers, an Indigenous community (Fort McKay First Nation), and institutions (including firms, electricity regulators, experts, and NGOs) with regards to potential solutions for decreasing the environmental impact of the Alberta oil sands. The pathways were developed from existing policies and studies that consulted a range of stakeholders.

The first pathway takes the government as the dominant narrator and was developed based on the Alberta Climate Leadership Plan. The Alberta Climate Change Advisory Panel was formed to provide insight from a broad group of stakeholders that include the Indigenous communities, industry, farmers, researchers, and the general public. Two public open houses were organised with input from over 1000 participants and an online survey (including questions using the Likert scale and open-ended questions) with over 25,000 respondents (Hill + Knowlton strategies, 2015; Leach et al., 2015).

The second pathway focuses on the local needs and interests of the Fort McKay community. The details for the pathway were inspired by a project commissioned by the Fort McKay Sustainability Office (Cumulative Effects Project) and carried out by the ALCES and Integral Ecology Group (ALCES and Integral Ecology Group, 2013). The results of the study explored the cumulative effects of oil sands development and proposed a paced approach to industrial development in their traditional territory. The report emphasised the community's interest in protecting the ecological integrity of an area of land while allowing its responsible industrial development and economic benefits for the society. This study identified specific ecological indicators the community was interested in studying since they observed drastic changes in the population (the availability of these species) that had directly impacted the community's way of life: moose, fisher, native fish, and edible berries. These indicators were not typically part of the government's indicators for climate change action, which primarily consists of CO_2 emissions. The views and thoughts presented for this pathway correspond to the authors' own analysis to the best of our knowledge, and do not reflect the formal stands of the Fort McKay First Nation or any of its members not listed as authors on this piece.

The third pathway, deployment of distributed renewables, is proposed by the Energy Futures Lab, a branch of The Natural Step Canada, an NGO that pushes an agenda to promote a sustainable transition in Alberta. The Energy Futures Lab is a platform that brings together a diverse range of partner organisations including industry, government, NGOs, and Indigenous communities. The proposed pathway is supported by energy associations, city economic development organisations, and electricity systems operators.

Aside from including existing policies and research initiatives, we have independently carried out 17 open-ended interviews with stakeholders over the period of November 2016 to April 2018: academics (four interviews), Indigenous community members (six interviews), industry players (five interviews), a non-profit organisation (one interview), and a policymaker (one interview).

We also carried out a risk elicitation by capturing the opinion of stakeholders during two public workshops. The first workshop was part of an existing conference, the Alberta Ecotrust Environmental Gathering. Researchers carried out a 'Solution forum' to explore how consensus can be reached within groups with various perspectives about sustainable development of natural resources in Alberta. The event took placed at Mount Royal University on 10 March 2018. There was a total of 24 participants. The attendees consisted of government bodies, politicians, industry members, researchers, and NGOs. Reponses were captured through small group role-playing exercises and a live polling.

The second workshop, 'Creating a common language for lower-carbon futures in Alberta', was organised within an existing seminar series run by the Graduate Colleges at the University of Calgary. The workshops consisted of a panel discussion. The interview panel including an industry representative, a policy consultant, an Elder from the Fort McKay community, a representative from the Fort McKay Sustainability Office, and an academic. The panel discussion took place on 12 March 2018.

Lower-carbon futures for the Alberta oil sands

This narrative considers three lower-carbon pathways for the Alberta energy sector including the Alberta oil sands: 'cap the emissions hat', developed as the government of Alberta's formal stance on capping emissions in the oil sands sector; 'hold your horses', which is inspired by the collective desired future to pace oil sand development of the Fort McKay community, representing over 775 Dene, Cree, and Métis people; and 'mix and round it all up', created from a combination of views that includes renewable and alternative energy while reducing emissions from the oil sands, as conveyed by the Energy Futures Lab. These different futures represent pathways that are not mutually exclusive and are partially complementary. They reflect alternatives developed under different contexts but aiming for a similar goal: to create a sustainable future for Alberta. These three futures will be presented by considering the views and expectations of stakeholders and the potential implementation and consequential risks of following a specific pathway.

'Cap the hat': carbon emissions cap for the oil sands and methane reduction

This future considers the government of Alberta's current effort to decrease emissions in the oil sands sector. As stated by Alberta's premier, leader of the New Democratic Party (NDP) responsible for development of the new policy for climate change abatement, Alberta is looking to develop a plan capable of decreasing the environmental impact of the oil sands sector while securing the economic benefits of that sector:

> Responding to climate change is about doing what's right for future generations of Albertans – protecting our jobs, health and the environment. It will help us access new markets for our energy products, and diversify our economy with renewable energy and energy efficiency technology. Alberta is showing leadership on one of the world's biggest problems, and doing our part.
>
> (Rachel Notley, Premier of Alberta, in Alberta Ministry of Environemnt and Parks, 2015)

As part of its Climate Leadership Plan, Alberta proposes an output-based allocation system for carbon emissions based on three main actions: (i) a 100 Mt emissions cap for the oil sands sector; (ii) establishing a credit trading system between emitters; and (iii) setting a C$30/ton carbon tax for facilities that exceed 100,000 CO_2 tons/year (Leach *et al.*, 2015). The carbon tax is expected to increase to C$50/ton by 2022 in order to meet federal targets (Government of Canada, 2016). With this plan, the government expects to reduce emissions by encouraging companies to implement new low-emission technologies. The industry is envisioned to continue growing, as the cost of production has decreased to around C$25/barrel due to technological efficiencies and will continue to decline in costs over forthcoming years (Erickson, 2018). At the current carbon tax rate, the cost of oil sands production has increased by C$1 (Ignjatovic, 2016). The carbon tax is expected to encourage industry to reduce emissions from bitumen extraction and production processes. Additionally, the carbon tax is expected to promote energy efficiency and the application of renewable energy in the extraction and production process, along with reducing methane flaring.

Aside from carbon emissions, targets have been set to decrease methane emissions by 45% by 2025. Methane cuts are expected to be achieved through setting emissions standards, improving monitoring, reporting, and verification of methane emissions, and detecting leaks and repairing (Leach *et al.*, 2015). In this future, carbon capture and storage (CCS) is anticipated as a game-changing technology. Starting from 2018, Alberta was expected to capture 2.76 million tons of CO_2 per year, equivalent to the emissions of around 600,000 cars/year (Alberta Ministry of Energy, 2017). We can anticipate the development of CCS to continue to grow once the cost of the technology falls. The Albertan government intends to lead the way towards a more sustainable energy system while respecting the current system that is largely dependent on fossil fuels. Technological advancements are expected to reduce emissions and environmental impact while increasing productivity. However, some technologies being tested, such as solvent-assisted in-situ bitumen extraction, although capable of generating a significant decrease in carbon emissions, raise many concerns due to their potential environmental impact and unknown cumulative effects in the local ecosystems. Alberta's lead is in line with federal intentions to reduce emissions through its Pan-Canadian Framework on Clean Growth and Climate Change. This framework contributes to Canada's 2017 National Determined Contribution to reduce emissions to 30% below 2005 levels by 2030.

The reception of this pathway among industry stakeholders is mixed. Some players consider that the Pan-Canadian framework could bring clarity and stability to the sector and potentially open new markets if the carbon footprint of the Alberta oil sands can be lowered through technological innovation. However, for other industry stakeholders, this future is harmful to the sector and lacks inclusion of the industry's current concerns. Although this future represents a change widely expected by many, the collapse of oil prices in 2014 changed the sector's outlook. Low oil prices, especially for the Alberta oil sands, forced producers to significantly cut spending and concentrate their efforts in reducing production costs to at least break even at oil prices below C$50/barrel (Erickson, 2018). Under this reality, the new cap on emissions, a trading system, and the carbon tax was an additional cost unwanted during a time when the industry was struggling. While the policy was being developed, oil producers increased production to maintain profits, causing an increase in GHG emissions (Environment and Climate Change Canada, 2017).

By contrast, these measures were welcomed by environmentalists, who also claimed the measures did not do enough to decrease climate change impacts. Moreover, this future represents a risk of double standards since the advisory panel that developed the Climate Leadership Plan acknowledged that such measures would probably not be enough to meet the emission reduction goals of the Paris Agreement and the Canadian NDC. Detractors of these new measures stress the fact that more-drastic changes are needed in order to achieve the global objectives. However, the limits placed of these regulatory actions were justified by avoiding carbon leakage – an increase of emissions in a jurisdiction as a result of the reduction of emissions in another jurisdiction with stricter climate policies. From our perspective, the actual reason was the desire to maintain an important sector and an economic system that is culturally embedded within the province's identity, since the majority of the oil sands reserves are present in Alberta. The Climate Leadership Plan considered in this pathway has been a political flag for the current ruling New Democratic Party (NDP) of Alberta. However, current economic challenges in the oil and gas sector have created strong opposition to this plan among various sectors in the province. Under a probable scenario of change in leadership by the next provincial elections in 2019, there is a high probability that alternative strategies for pollution control are implemented instead of the Climate Leadership Plan. For example, the United Conservative Party have promised to eliminate the NDP's plan if elected. Such political uncertainty within this pathway can jeopardise the possibilities of Canada to meets its NDC goals.

'Hold you horses': paced oil sands development and land-use protection

This future was developed inspired by the views of the Fort McKay First Nation which, seeking advice from scientists and their western perspectives, developed a plan to protect their land's ecological integrity and their traditional ways of living while offering an alternative to maintain the industrial sector operating in the area (ALCES and Integral Ecology Group, 2013). The perspectives presented on this pathway were developed by the authors to the best of their knowledge and do not represent the official stand of the Fort McKay First Nation or any of its members not listed as chapter authors.

This pathway considers an approach to have the government of Alberta manage the pace of bitumen production within the Athabasca oil sands area, comprising of the 3.62 million (M) hectares. The size of the Athabasca oil sands area is equivalent to nearly 10% of the provincial-owned forest coverage. There is a strong emphasis on maintaining traditional land uses and protecting wildlife while enabling the oil sand sector to develop at a more thoughtful pace. This pathway suggests several implementation strategies. A key strategy is to expand protected areas and maintain levels of ecological disturbance below a threshold that allows the recovery of the area. This study suggested that increasing protected areas and managing industrial activity in between 10.4% and 39.2% (378,483–1,420,579 ha) of the Fort McKay's Traditional Territory could improve the ecological integrity and biodiversity of the area (ALCES and Integral Ecology Group, 2013). The Traditional Territory is the land entitled to the people of Fort McKay to exercise their treaty rights as stipulated in Treaty 8, which upholds the community's right to hunt, trap, and gather resources on the land (Canada, Privy Council, O.C. No. 2749, 1966). According to the words of Selina Harpe, a Fort McKay Community Elder:

> The land is our livelihood, we live off this land all around here. We grew up in the trap-line, and at younger age my dad did trapping and hunting, there was no other way to make a living. At that time there was only native

people in town, before Europeans came, so everyone did whatever they had to do to make a living for your family, and that is the way we used to live.

(quoted in Behr, 2013)

The proposed resources management plan contemplates: the establishment of land disturbance limits; defining and implementing culturally relevant reclamation plans for damaged areas; strict monitoring, reporting and compliance to specific air quality standards; controlled surface and ground water withdrawals and effluent quality monitoring; protection and reclamation of wetlands; ensuring fish and wildlife impact mitigation and monitoring; and protecting traditional routes and areas located in the zone (Alberta Ministry of Environemnt and Parks, 2018a). The protected land areas can be aligned with the 2012 federal policy *Federal Recovery Strategy for Woodland Caribou – Boreal Population* which recommended that implementation plans should be set at the provincial level to protect 65% of caribou ranges by October 2017 (Environment Canada, 2012). Currently in Alberta, between 57% and 95% of each caribou range is disturbed by industrial activities (Russell, Pendlebury, and Ronson, 2016). Other important strategies are to demand rigorous industrial best practices for companies operating in the area to reduce land and water use, and to increase the footprint reclamation rate while considering biodiversity and potential future uses of the reclaimed land (ALCES and Integral Ecology Group, 2013). The implementation of this plan is expected to mitigate the effects of industrial development and enable the recovery of the ecology and biodiversity of the area. These changes could allow the members of the community of Fort McKay to practice their traditional way of living and knowledge dissemination practices. In addition, this plan is expected to still allow significant bitumen extraction and related industrial activities in order to maintain the influx of economic benefits from the traditional territory.

The intention behind land management is to provide a protected space for the ecosystem and to support traditional land-use practices. Parts of the land have been used by families for hunting and trapping for many generations, while berries and herbs are collected for consumption and medicinal uses. Securing these ranges of land provides assurance that both environmental and traditional practices can be preserved while the industrial activity keeps its course. There are also hopes that a paced development might lead to more thoughtful resource-extraction practices. The communities living close to the oil sands would like to be able to drink tap water without the fear of health concerns from the effluent released by industry. They would also like to keep their windows open to let in the breeze without concerns of foul smells and poor air quality (Thurton, 2018). If the land around the community is allowed to recover, community members can once again pick berries and consume them without the fear of contamination from the land. The moose and birds may also return to the land and bring with them the assurance that the ecosystem is being restored. The protection of areas of land is expected to have a positive impact on the health of ecosystems needed for species to maintain their population. Members of the community are not only calling for the right to carry out their practices, but also for the fundamental right to a clean living environment:

> We need and have the right to clean air and water, and enough healthy plants and animals to harvest in need. If Alberta and the industry change the ways things are done, we can have this things and Oil Sands can continue to be develop.
>
> (Holly Fortier, Fort McKay cross-cultural consultant and community member, quoted in Behr, 2013)

Based on a land-use model, the ALCES Group, a land-use consulting firm, was able to identify mitigation options supported by the community that, if incorporated into provincial regulations and implemented, could allow the recovery of the moose habitat and increase the livelihood of other animal and plant species, while allowing the economic benefits of continued bitumen extraction in the Fort McKay Traditional Territory. The most unique factors for this study were the biotic indicators appointed by the community as essential for the wellbeing of the area (moose, fisher, fish, edible berries, and others), and the critical areas to protect based on the areas traditionally used for hunting, trapping, fishing, and spiritual traditions. This pathway also implies strong collaboration between industry and the First Nations communities most directly impacted by bitumen production. Industry will need to collaborate to set best practices and to consult with communities in the expansion of oil sands extraction and reclamation of land. Government support would also be required to ensure that the land-use designation is respected by industry. The paced development should increase the quality of air and reduce the amount of water extracted for industrial use.

Despite of the efforts of the Fort McKay community to implement this plan, since 2013 no definitive action has been seen from the government. According to court records, several attempts to implement this plan have been carried out with no success. On several occasions, Fort McKay had to pursue legal routes to stop development projects and promote a paced development, as evidenced in the four legal cases Fort McKay has conducted since 2013. More recently, Fort McKay has been involved in legal actions to demand the management of resource development near their reserve lands. To date, some authorities have heard the community's proposals and have considered developing a plan for the area they want to be protected. Shannon Phillips, Minister of Environment and Parks, has publicly announced the consultation with the community, indicating:

> We are moving forward in collaboration with the Fort McKay First Nation to protect their traditional lands, while ensuring responsible development near their community. By working jointly with Indigenous peoples and industry, we have made a lot of progress and I look forward to refining the plan after careful consultation.
>
> (Shannon Phillips, Minister of Environment and Parks, in Alberta Ministry of Environment and Parks, 2018b)

This pathway – although inspired by the perspectives of one Indigenous community defined by their particular values, traditional territory, and unique relationship with the oil industry – could be extrapolated to the areas of Alberta where the interests of industrial activities and the protection of ecosystems are in conflict. Parting from the knowledge and values of local communities, comprehensive climate change action and mitigation plans that allow responsible industrial activities could be developed with high chances of success, if implemented. However, political changes, resistance from industry players to a slower growth, the misunderstanding and misperceptions of the cultural values of the community, and potential economic and job losses could stop this future from being fully implemented.

'Mix and round it all up': supporting a clean energy mix

This future is based on the innovations pathways proposed by the Energy Futures Lab, which looks into leveraging Alberta's strengths to transition towards a sustainable energy system. It represents collaboration among a range of stakeholders with vested interests in the energy sector including: fellows from the energy industry; First Nations and Métis communities; the provincial government; and environmental organisations. It also includes targets set by the Climate Leadership Plan. The pathway considers limited growth in the oil sands sector and the expansion of renewable energy, with an aim of increasing renewable energy to 30% by 2030 in the electricity sector. This implies limiting the development of existing oil sands sites and replacing Alberta's coal-generation capacity with two-thirds from natural gas and one-third from renewable energy (adding 5000 megawatts by 2030 by the Renewable Electricity Program (REP)) (Alberta Electric System Operator, 2018a, 2018b). The development of renewable energy is estimated to bring in C$10.5 billion in new investment by 2030 and potentially create more than 7,200 new jobs. It will be rolled out through a competitive bidding process (Audette-Longo, 2017). The first bidding round is expected to lead to installation of 400 megawatts of renewable electricity generation capacity.

This pathway considers increasing the carbon tax (greater than C$30/ton) to level the playing field with fossil fuels production. In order to meet energy shortfalls, the pathway will explore the development of renewable energy, primarily solar power and wind power, and the use of geothermal energy produced from over 400,000 wells in Alberta.

Government-initiated programmes can help with the transition to renewable energy by providing funding and subsidies. Particular support will be provided to the 48 First Nations communities in Alberta: C$35 million has been invested to support local renewable energy projects, encourage energy efficiency, and provide training for jobs in the low-carbon economy. Additionally, renewable energy development can be supported by recycling revenues from the carbon tax amounting to C$5.4 billion over a three-year period. Revenues from the carbon tax can be directed to those most adversely impacted by higher energy prices. Around 27%, or C$1.5 billion, will be directed as rebates to lower- and middle-income Albertans to offset the carbon tax.

The future of this pathway is anticipated to provide more meaningful employment, particularly for those transitioning out of the oil sector and for young graduates in their early twenties entering into the labour market. There is also significant potential for renewable resources. Alberta can be seen as a technological innovator for energy systems by tapping into the abandoned gas wells currently viewed as a liability for geothermal energy. There is also potential for decentralised wind and solar power systems, primarily where resources are more abundant in the more populated southern regions. Solar systems can be installed in homes, while wind farms can be located on the outskirts of the city (but not too far from the electricity demand). In essence, this future considers that natural resources, including metals and fossil fuels, are used in a sustainable enough manner to allow a thriving human population capable of maintaining an equilibrium with the environment. In the words of Karl-Henrik Robert in his book *The Natural Step Story: Seeding a Quiet Revolution* (Robert, 2002):

> In the sustainable society, flows of matter are balanced or, at least, not systematically unbalanced. Natural cycles surround society and define the limits within which we have to live. The sustainable society lives partly on flows from nature's production and partly on smaller flows of metals and minerals from the Earth's crust. Plants build up enough renewable resources to satisfy their consumption by animals and humans.... Since (in the sustainable society) the rate of this flow does not exceed the rate of regeneration of resources, it can be regarded as an 'interest rate' from nature rather than as a systematic toll from its 'capital'.

Some aspects of this future are already being implemented under the Alberta government and the Climate Leadership Plan. However, some elements remain at a conceptual level. The Energy Futures Lab lacks specific action plans that are made available to the general public. Instead, it is currently being implemented by leveraging the connections and influence of its fellows to help educate professionals in roles capable of catalysing change in the energy sector. However, despite being co-developed by a diverse group of people, this approach may lack a perspective on the communities most impacted by energy development activities, for which these changes could bring the most benefits or disadvantages. This could imply a significant challenge for implementation of this vision. Other barriers could be the significant change of local policies to stimulate the development of renewable energy transformations at both individual and industry levels. Alberta, as of April 2018, approved a bill looking into a Property Assessed Clean Energy (PACE) programme that allows property owners to install renewable energy capabilities in their facilities financed by property taxes, providing financial incentives for renewables development. However, it will be up to the municipalities to decide if they want to take part in this programme, hindering opportunities for interested individuals if their municipality opts to not participate in the programme. Another initiative taking place is the request of the Alberta Electric Systems Operator (AESO) to incorporate 30% renewable generation into the grid by 2030. The first call was a success in achieving an electricity cost offer of 3.7 cents per kilowatt, the lowest renewable electricity pricing in Canada to date, and it is expected to become operational by 2019 (Government of Alberta, 2018a, 2018b).

The best of all worlds

From consultation, conversations, and discussions, the desire of most Albertans is to have the best of all pathways. The people of Alberta, across all sectors, seem to want a growing energy business based on fossil resources like the oil sands while decreasing the environmental impact of the sector in the territory. Some may argue this contrasting and idealistic vision might be impossible to materialise. However, if stakeholders are convinced this is a way forward, then the main challenge remaining is developing a process to communicate different worldviews to co-develop the mechanisms that could make such a vision come true. Such understanding of each other's realities is vital to achieve consensus. Today, the lack of consensus represents the biggest barrier to build a feasible lower-carbon pathway for the Alberta oil sands.

Public perception collected from interviews, surveys, and media indicate Canadians are committed – in principle – to climate change action. This commitment may take different forms nationally or regionally. In 2014, 65% of Canadians surveyed claimed to be familiar with the discussion around the use of fossil fuels and their impact on climate change, and 62% agreed that protecting the environment was more important than the price of energy, indicating a preference for government measures to implement taxes on the main polluters (Nanos Research, 2014) seen in the prairies region where Alberta is located. In the same year, another survey revealed that approximately 43% of the population was completely or partially in favour of oil sands projects in Alberta, mainly due to their economic and employment benefits. However, a growing 35% indicated that their impression of the oil sands projects has worsened over the past five years.

In 2015, a survey performed by a major broadcasting company (CTV) indicated that 72% of Canadians believed that the science of climate change is irrefutable. In addition, 63% of Canadians were willing to pay more for certain products in order to help the country meet its environmental commitments, despite the fact that 66% (45.5% in the Alberta region) acknowledged they were aware that meeting the new target may involve a significant job loss in the Canadian oil industry (CTV and Nanos Research, 2015). Considering the public's opinion, the government carried out a national consultation to define a provincial operating budget for 2016, receiving significant feedback regarding the use of green technologies and the growth of the green energy sector. This feedback helped the government to outline a comprehensive strategy for the sector. In 2016, a survey revealed the views of the Canadian public on climate change initiatives. In this survey, more than three out of four Canadians supported a national plan that guaranteed meeting international targets for emissions reduction, supported independent action from the federal government to meet the targets, and had a positive perspective on establishing a minimum carbon pricing scheme across Canada (Nanos Research, 2016).

Based on stakeholder consultations carried out in Alberta (see the *Research process and methods* box earlier in this chapter for details), a dominant view emerged that was favoured by a range of Alberta stakeholders. Stakeholders would favour a future that converged the development of the oil sands sector with the growth of energy supply from renewable resources (Figure 4.2). Such a vision was observed across all the stakeholders consulted. Stakeholders from industry, government, environmental groups, communities, and Indigenous communities all considered a future where oil sands continued being one of the economic drivers of the province. The graduate population from local universities are now seeking more meaningful employment morally aligned with their own values where they can contribute to the greater good of society. More Albertans are becoming aware of both the benefits, as well as the negative impacts, of the oil sands. While there is an agreement that a gradual transition is needed, there is greater hope placed on cleaner technologies that support a more sustainable and equitable future. Such a middle-ground position indicates that Albertans hope for a solution able to keep the best of all worlds: maintain an economic benefit from a growing oil and gas sector while assuring the protection of the ecological integrity of the territory. This preference, although idealistic, resonates with elements present on the 'mix and round it all up' future, which supports a clean energy mix incorporated into the current energy supply.

Indeed, ideals for the hopes of Albertans as a whole, the development of the oil and gas sector, and even the development of the renewable energy sector will continue to have negative impacts on the environment unless new technologies are introduced that are capable of changing current practices towards unknown, low-carbon economic activities. Moreover, communities where these activities occur will not be where most of the population concentrates. Instead they are likely to be located near remote communities, with high probabilities of conflicting with Indigenous values and traditional practices. Still, with the implementation of a cleaner energy mix, local communities will pay the price to maintain a more sustainable – better publicly perceived – Alberta: the sacrifice of their surrounding ecosystems and wellbeing due to the deployment of energy generation facilities strange to the local ecosystem. Considering these aspects, we asked Albertans to identify the biggest risks they foresaw for this ideal future (Figure 4.3).

All the risks identified by stakeholders could be grouped in six main risks classified by their effect on the pathway, either as a barrier (implementation risk) or as a negative impact (consequential risk). The implementation risks include: (i) social rejection, where public opposition may affect the realisation of such a pathway because of anticipated impacts; and (ii) political barriers,

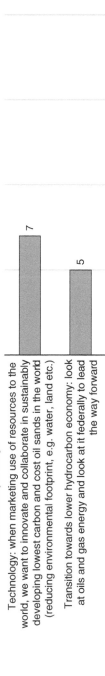

What kind of future do you foresee/envision for the energy sector in Alberta in terms of development and sustainability?

Low-carbon pathways

Technology: when marketing use of resources to the world, we want to innovate and collaborate in sustainably developing lowest carbon and cost oil sands in the world (reducing environmental footprint, e.g. water, land etc.) — 7

Transition towards lower hydrocarbon economy: look at oils and gas energy and look at it federally to lead the way forward — 5

Converging pathways: renewables (continued growth) and optimising existing energy systems around hydrocarbons for greater efficiency and moving to higher valued added products in the most environmentally and more connected ways with other provinces. — 19

Need to look at a Plan B beyond the current oil development plan; with industry improving the environmental issues need to go beyond green house gases (e.g. tailing ponds) — 7

Need industry to have environmentalists to work within each department in industry that sincerely care about the environment and community; looking toward the future for the next generation — 7

Answer count (Total results: 45)

Figure 4.2 Preferred low-carbon future by surveyed stakeholders.

Source: workshop #2, University of Calgary, 12 March 2018.

Figure 4.3 Preliminary risk assessment developed by surveyed stakeholders.

Source: workshop #1, Alberta Ecotrust Environmental Gathering, Mount Royal University, 10 March 2018.

especially the weak integration of federal and provincial legislation negatively affecting implementation of the described pathway. There are more consequential risks identified in four broad categories. First, negative environmental impacts – such as the pathway may still cause damage to local ecosystems due to resource extraction, land-use change, air and water pollution, the creation of physical barriers, and noise pollution. Second, negative health impacts, particularly industrial pollution, which may affect the health of local communities if this pathway were to be pursued. Third, negative impacts on local communities, such as effects on the traditional ways of living of Indigenous communities. And fourth, negative economic impacts, especially a potential drop in profits leading to job losses and contraction of the economy. The dynamic between the risks and uncertainties found shows the complex context given for the development of lower-carbon pathways for the Alberta oil sands. In addition, the interconnectivity between all factors suggest that policy development should be done evaluating the sector as a system, and not focusing efforts on isolated approaches that could negatively impact the effectiveness of the much-needed policies (Figure 4.4).

Surveyed stakeholders in Alberta are aware of the challenges faced by interested parties – government, industry, and Indigenous communities – when trying to find common ground. Even under ideal conditions – a future where oil and gas become sustainable – Albertans identify conflicts inherent to the interests of the defined institutions. For industry, economic losses play a role,

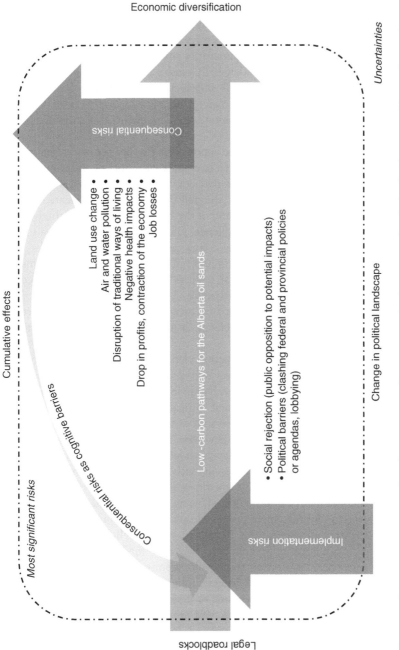

Figure 4.4 Main uncertainties and implementation and consequential risks associated with the development of low-carbon pathways for the Alberta oil sands.

Economic diversification

Consequential risks

Land use change •
Air and water pollution •
Disruption of traditional ways of living •
Negative health impacts •
Drop in profits, contraction of the economy •
Job losses •

Low-carbon pathways for the Alberta oil sands

• Social rejection (public opposition to potential impacts)
• Political barriers (clashing federal and provincial policies or agendas, lobbying)

Uncertainties

Change in political landscape

Cumulative effects

Consequential risks as cognitive barriers

Most significant risks

Implementation risks

Legal roadblocks

while for the government social rejection and political instability were identified as the main risks. Moreover, when asked to reflect about the issues from the local community perspective, participants did not have problems identifying the issue of continuing damage to ecosystems, leading to health impacts and disruption of the traditional Indigenous ways of living. These risks reflect some of those claimed by Indigenous communities since the arrival of Europeans to the Americas. So, we might wonder, is this just a matter of a lack of efficient strategies, or does this reality reveal a deeper conflict? Is the absence of a willingness to step out of its own defensive zone and address this issue from the initial intention to find a common ground the core element of a dysfunctional conversation?

Such questions do not consider the approach of the Indigenous communities since historically they have been forced to comply with unfair conditions and to adapt to European cultural principles that conflict with their own worldviews. We argue that this question seeks the reflection of those educated under Western values that lack understanding of the holistic existence of the human as part of the ecosystem. As a consequence, finding the proper language to use in the different contexts could be a key practice to minimise risks and uncertainties in climate change action and mitigation. This could help ensure that plans are able to adapt to local realities and have a greater chance of successfully protecting vulnerable ecosystems. Maybe this is the real 'best of all worlds': finding a way to co-develop, by means of Indigenous and Western worldviews, an inclusive plan capable of decreasing the impacts of climate change.

For an oil sands worker, a climate action plan could be presented as a need to adapt the sector to changing markets and decrease liabilities in a changing investment landscape. On the other hand, for a member of an Indigenous community, climate change action and mitigation could be presented as the opportunity to save moose and fish, as well as to guarantee the permanence of healthy land available for traditional practices. Therefore, in order to develop a successful action plan, multiple roadmaps should be adapted to local realities. There is as yet no perfect route to achieving such changes, and maybe there never will be. However, a future that considers the local context and co-development of action plans could be a starting point, allowing for implementation of effective plans to transition towards a more sustainable and equitable future.

When looking at the futures identified, all stakeholders claimed that participation of Indigenous stakeholders was encouraged, and that feedback was taken into consideration on the development of the futures 'cap the hat' and 'mix and round it all up'. However, Indigenous values are not evidently reflected in the language, action plans, or main interpretation of such action plans. These futures were developed, and are expected to be measured and implemented, from a colonising perspective. In contrast, the future 'hold your horses' was developed starting from Indigenous values of land and ecosystems, and then seeking Western scientific principles to translate its benefits to current Western understanding. This future portrays a social–ecological dynamic with actors – e.g. humans, community, animals, plants, water – on a level playing field. The definition of ecological integrity from a perspective of biotic indicators (moose, fisher, fish, edible berries' habitats), demonstrate a local interpretation of environmental impact which stakeholders can relate to, are able to monitor, and can commit to protecting. As reported by Whitmarsh, Seyfang, and O'Neill (2011), community involvement is a key factor for the success of policies for climate change mitigation. Therefore, from the local context, the pathway 'hold your horses' is expected to be more successful in achieving its goals – if implemented – rather than the other futures that lack elements that local communities can be represented in and be relate to. This scenario could be extrapolated to the national plans for climate change mitigation and adaptation. Today, the international discourse is defined on GHG emissions, a concept that is not tangible or relatable to challenges present at the local reality of the communities.

Future outlook

In the quest for identifying lower-carbon pathways envisioned by the stakeholders in the Alberta oil sands sector, three main scenarios were identified: one following the official proposal from the provincial government, one based on Indigenous needs and priorities, and one based on a combined strategy created by the collaboration of several groups and aligned with federal strategy. Within these pathways, implementation and consequential barriers were captured from stakeholders as potential risks should this future materialise. These barriers mostly concentrated around economic losses, environmental degradation, social imbalance, and health impacts. Whether these risks were noticeable depended on the point of view of each stakeholder. However, a bigger risk is linked to the lack of consensus among stakeholders and the absence of a pathway capable of unifying and representing all viewpoints.

From the futures captured, the Indigenous communities most directly impacted by the oils sands development, i.e. where the sector develops in traditional land in the vicinity of their communities, were the least represented among all. Although efforts to consult and include Indigenous communities has been made for all narratives, true inclusion or representation is not evident in their actions plans. This means that any action or policy towards reducing emissions in the oil sands sector will have considerable impact on the welfare of the neighbouring Indigenous communities, who are left with no voice as to their fate. Such realities call for urgent action in incorporating Indigenous/local communities in the process of co-developing policies effective to meet the commitments of Canada in the Paris Agreement. As a way forward, a consensus-building framework could serve as an important tool to maintain the interest and commitment of the participants. One example is the Consensus Building Engagement Process (CBEP), focused on step-by-step consensus building through the consultation and inclusion of all affected stakeholders (Lieu *et al.*, 2018). The framework could consider Indigenous needs, priorities, and knowledge and by build consensus in the planning and execution of the project/initiative. Understanding local values and allowing local communities to have an active role in the development process could generate benefits in the area of effectiveness, social support, and innovative ways to promote climate action.

The study of the multiple pathways shared by stakeholders in the Alberta oil sands highlights the need for developing effective climate action capable of incorporating and representing local interests and concerns. Although efforts has been made in the past to improve representation of several sectors, evenly levelled participation through incorporation and consensus could be more effective that simple inclusion. A process to consider local communities as equal participants in the co-development process of policy could be more effective than those changes implemented so far, which have attracted plenty of criticisms for their lack of community representation. There is no formula for this approach since such action would require acknowledging varying cultures and worldviews. However, if policies are developed and/or implemented with this notion, the gap between climate policy and real on-the-ground impact could be reduced.

References

Alberta Electric System Operator (2018a). *Renewable Electricity Program*. Available at: www.aeso.ca/market/renewable-electricity-program/ [accessed 7 June 2018].

Alberta Electric System Operator (2018b). *Renewable energy to power new jobs, investment*. Available at: www.alberta.ca/release.cfm?xID=55695ADD3C996-CDCD-89D9-6A025089F93B3584 [accessed 7 June 2018].

Alberta Energy Regulator (2015). *ST98–2015: Alberta's Energy Reserves 2014 and Supply/Demand Outlook 2015–2024*.

Alberta Ministry of Culture and Tourism (2016). *Oil Sands*. Available at: www.history.alberta.ca/energyheritage/sands/default.aspx.

Alberta Ministry of Energy (2016). Energy's History in Alberta. Available at: www.energy.alberta.ca/About_Us/1133.asp.

Alberta Ministry of Energy (2017). *Carbon Capture and Storage (CCS) Fact Sheet*. Available at: www.energy.alberta.ca/AU/CCS/Pages/default.aspx.

Alberta Ministry of Environment and Parks (2015). *Climate Leadership Plan will protect Albertans' health, environment and economy*. Available at: www.alberta.ca/release.cfm?xID=38885E74F7B63-A62D-D1D2-E7BCF6A98D616C09.

Alberta Ministry of Environment and Parks. (2018a). *Draft Moose Lake 10 km Management Zone Plan (Under Review)*. Available at: https://talkaep.alberta.ca/draft-moose-lake-10-km-management-zone-plan.

Alberta Ministry of Environment and Parks (2018b). *Draft Moose Lake Management Zone Plan released*. Available at: www.alberta.ca/release.cfm?xID=52421444A7C71-BB61-CD7A-621D064F4B04181C [accessed 7 June 2018].

ALCES and Integral Ecology Group (2013). *Fort McKay Cumulative Effects Project Technical Report of Scenario Modeling Analyses with Prepared for the Energy Resources Conservation Board on behalf of the Fort McKay Sustainability Department* (1673682).

Audette-Longo, T. (2017). *Wind energy jobs, cash could boost Alberta's economy*. *Canada's National Observer*. Available at: www.nationalobserver.com/2017/10/04/news/wind-energy-jobs-cash-could-boost-albertas-economy [accessed 7 June 2018].

Behr, T. (2013). *Moose Lake: Home and Refuge*. Fort McKay, Canada. Available at: www.fortmckayvirtualmuseum.com/.

Canada, Privy Council, O.C. No. 2749 (1966). *Treaty No. 8*. 6 December 1898. Department of Indian Affairs and Northern Development (DIAND), Ottawa.

Canadian Association of Petroleum Producers (CAPP) (2016). *Oil Sands History and Milestones*. Available at: www.canadasoilsands.ca/en/what-are-the-oil-sands/oil-sands-history-and-milestones.

Chastko, P. (2000). *Developing Alberta's Oil Sands: From Karl Clark to Kyoto*. University of Calgary Press, Calgary. Available at: http://site.ebrary.com/lib/ucalgary/docDetail.action?docID=10134862.

CTV and Nanos Research (2015). *Canadians say climate change a threat to Canada's economic future and that Canada's reputation has taken a hit on the global stage*. Available at: www.nanos.co/our-insight/.

Environment Canada (2012). *Recovery Strategy for the Woodland Caribou (Rangifer tarandus caribou), Boreal population, in Canada. Species at Risk Act Recovery Strategy Series*. Ottawa. Available at: www.registrelep-sararegistry.gc.ca/default.asp?lang=En&n=33FF100B-1.

Environment and Climate Change Canada (2017). *Canadian Environmental Sustainability Indicators: Greenhouse Gas Emissions*. Available at: www.ec.gc.ca/indicateurs-indicators/default.asp?lang=En&n=FBF8455E-1.

Erickson, P. (2018). *Confronting carbon lock-in: Canada's oil sands. SEI discussion brief*. Seattle, WA. Available at: www.sei.org/publications/confronting-carbon-lock-canadas-oil-sands/.

Finkel, A. (2000). *The Social Credit Phenomenon in Alberta*. University of Toronto Press, Toronto. Available at: http://site.ebrary.com/lib/ucalgary/docDetail.action?docID=10200948.

Government of Alberta (2018a). *Property Assessed Clean Energy (PACE) legislation*. Available at: www.alberta.ca/PACE.aspx [accessed 7 June 2018].

Government of Alberta (2018b). *Renewable Electricity Program*. Available at: www.alberta.ca/renewable-electricity-program.aspx [accessed 7 June 2018].

Government of Canada (2016). *Pan-Canadian Approach to Pricing Carbon Pollution*. Available at: http://news.gc.ca/web/article-en.do?mthd=advSrch&crtr.mnthndVl=&crtr.mnthStrtVl=&crtr.page=2&nid=1132169&crtr.yrndVl=&crtr.kw=carbon+pricing&crtr.yrStrtVl=&crtr.dyStrtVl=&crtr.dyndVl=.

Government of Canada (2017). *Canada's 2017 Nationally Determined Contribution submission to the United Nations Framework Convention on Climate Change*.

Harner, T., Rauert, C., Muir, D., and 26 others (2018). Air Synthesis Review: Polycyclic Aromatic Compounds in the Oil Sands Region. *Environmental Reviews*. NRC Research Press. doi: 10.1139/er-2018-0039.

Hill + Knowlton strategies (2015). *ALBERTA CLIMATE LEADERSHIP DISCUSSIONS. Report from Online Engagement*. Available at: www.alberta.ca/documents/climate/CLD_Report_on_Engagement_-_Final_(Nov._17_2015).pdf.

Ignjatovic, D. (2016). *What Does the Carbon Tax Mean for the Canadian Oil Sands? Special Report*. Available at: https://economics.td.com/carbon-tax-canadian-oil-sands.

Institute for Oil Sands Innovation (2016). *Oil Sands History and Development*. University of Alberta. Available at: www.iosi.ualberta.ca/en/OilSands.aspx.

International Energy Agency (2015). *Energy Policies of IEA Countries: Canada 2015 Review*.

Leach, A., Adams, A., Cairns, S., Coady, L., Lambert, G. (2015). *Executive Summary. CLIMATE LEADERSHIP. Report to Minister*. Alberta Minister of Environment and Parks. Available at: www.alberta.ca/climate-leadership-plan.aspx.

Lieu, J., Virla, L.D., Abel, R., Fitzpatrick, C. (2018). 'Consensus Building in Engagement Processes' for reducing risks in developing sustainable pathways: Indigenous interest as core elements of engagement, in Doukas, H., Flamos, A., Lieu, J. (eds) *Understanding risks and uncertainties in energy and climate policy: Multidisciplinary methods and tools towards a low carbon society.* Springer.

McLachlan, S. (2014). *Water is a living thing: environmental and human health implications of the Athabasca Oil Sands for the Mikisew Cree First Nation and Athabasca Chipewyan First Nation in Northern Alberta.* Available at: https://landuse.alberta.ca/Forms and Applications/RFR_ACFN Reply to Crown Submission 6 – TabD11 Report_2014-08_PUBLIC.pdf.

Nanos Research (2014). *Oil Sands Summary.* Available at: www.nanos.co/wp-content/uploads/2017/07/2014-612-Oil-Sands-Report-with-tabs-R.pdf.

Nanos Research (2016). *Clean Energy Survey Summary.* Available at: www.nanos.co/wp-content/uploads/2017/07/2016-906-Clean-Energy-Populated-Report-Part1-w-tabs-R.pdf (Part One) and www.nanos.co/wp-content/uploads/2017/07/2016-906-Clean-Energy-Populated-Report-Part2-w-tabs-R.pdf (Part Two).

Natural Resources Canada (2018). *Energy and the economy.* Available at: www.nrcan.gc.ca/energy/facts/energy-economy/20062#L6.

Natural Resources Canada (2016). *Energy Fact Book.* Available at: www.nrcan.gc.ca/sites/www.nrcan.gc.ca/files/energy/pdf/EnergyFactBook_2016_17_En.pdf.

Parajulee, A., Wania, F. (2014). Evaluating officially reported polycyclic aromatic hydrocarbon emissions in the Athabasca oil sands region with a multimedia fate model. *Proceedings of the National Academy of Sciences* 111(9), 3344 LP-3349. Available at: www.pnas.org/content/111/9/3344.abstract.

Regional Aquatics Monitoring Program (2016). *History of the Oil Sands.* Available at: www.ramp-alberta.org/resources/development/mining.aspx.

Robert, K.-H. (2002). *The Natural Step Story: Seeding a Quiet Revolution.* New Society Publishers, Gabriola Island, B.C.

Russell, T., Pendlebury, D., Ronson, A. (2016). *Alberta's Caribou: A Guide to Range Planning Vol. 3: Northwest Alberta.* Edmonton. Available at: https://cpawsnab.org/wp-content/uploads/2018/03/CPAWS_Guide_to_Caribou_Range_Planning_Vol_3_NW_Herds_FINAL.pdf.

Tenenbaum, D.J. (2009). Oil Sands Development: A Health Risk Worth Taking? *Environmental Health Perspectives* 117(4), A150–A156. Available at: www.ncbi.nlm.nih.gov/pmc/articles/PMC2679626/.

Thurton, D. (2018). '*Smells like strong cat pee': Smartphone app tracks oilsands odour complaints.* CBC News. Available at: www.cbc.ca/news/canada/edmonton/wood-buffalo-air-quality-app-1.4703417 [accessed 7 April 2018].

Whitmarsh, L., Seyfang, G., O'Neill, S. (2011). Public engagement with carbon and climate change: To what extent is the public 'carbon capable'? *Global Environmental Change* 21(1), 56–65. doi: https://doi.org/10.1016/j.gloenvcha.2010.07.011.

5 United Kingdom

Pathways towards a low-carbon electricity system – nuclear expansion versus nuclear phase out

Rocío Alvarez-Tinoco, Michele Stua, and Gordon MacKerron

Introduction

Since the early 2000s, fossil fuel has registered a sharp decrease in its role in overall electricity generation in the UK, falling from more than 250 TWh (terawatt hours) supplied in 2005 to less than 150 TWh in 2015. With nuclear power holding its earlier share, renewables were the main source of substitution for fossil-based generation. Identifying the drivers for this shift in UK electricity supply represents the first step to understand the strategies, policies, and actions characterising the UK electricity generation system.

While prices and energy security concerns have maintained a strong influence on the UK electricity system, one more recent driver has been significant in determining major changes in the UK electric generation mix. The commitment in the UK Climate Change Act 2008 (HM Government, 2008) to achieve 80% emission reductions by 2050 has significantly raised attention to the potential for low-carbon electricity. An ageing electricity infrastructure, requiring significant renewal in the next decades, represents an opportunity to accelerate the UK's low-carbon electricity transition. Consequently, UK climate strategies, policies, and actions are likely to become the key drivers for the evolutionary dynamics of the UK electricity mix over the next decades.

Two opposing decarbonisation pathways are particularly relevant for the UK electricity system. One relies on an increase in new nuclear power and the other involves phasing out nuclear with increased reliance on renewable energy.

Comparison of the electricity mix over the past 35 years with the changes in the UK electricity generation mix between 2015 and 2016 (Figures 5.1 and 5.2) offers insights into the UK's decarbonisation process as a driver in transforming domestic electricity generation.

Apart from the significant shift from traditional fossil fuels to renewable resources, this comparison illustrates the dynamics within the British fossil-fuel sector. The dramatic decrease in the share of coal by more than 50% in just one year was not compensated by a rise in the low-carbon share, but by natural gas. While gas is a fossil fuel, it is often considered a 'transition' resource in a long-term

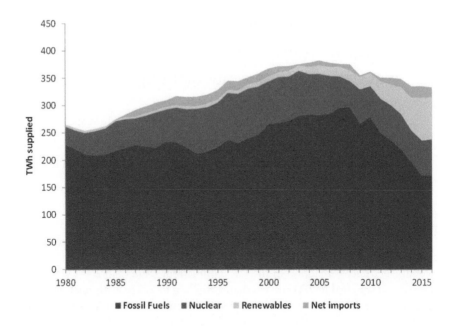

Figure 5.1 The dynamics of UK electricity supply, 1980–2015.
Source: DUKES, 2017, p. 113.

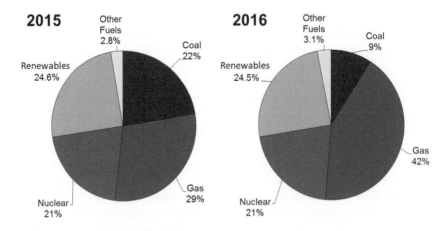

Figure 5.2 Share of electricity generation in the UK by resource, 2015–2016.
Source: DUKES, 2017, p. 117.

decarbonisation strategy due to its relatively low climate impact compared with coal and oil, as well as its relatively low price. However, if these dynamics are maintained in the short to medium term, major changes will be needed to meet long-term targets. Natural gas is incompatible with a successful decarbonisation process, although it is imaginable that gas with carbon capture and storage (CCS) could help decarbonise the electricity system. We have excluded it from this research. This is for two reasons: it is not clear that CCS will become commercially available; and none of the stakeholders at our consultation meetings raised CCS as an important low-carbon option. In addition, large-scale continued use of gas raises uncertainties for long-term energy security. With the UK government also aiming for a complete phase out of coal by 2025, new nuclear and renewable resources become central for genuine decarbonisation in the UK electricity system. Reductions in energy demand would also help facilitate this decarbonisation process by reducing the amount of new generating investment required. However, consideration of energy demand reduction is beyond the scope of this chapter.

Nuclear power policy

Nuclear power is a low-carbon electricity technology, able to contribute to electricity supply on a potentially large scale. However, nuclear power involves multiple risks such as safety (including cyber-security), radioactive waste management, and potential nuclear weapons proliferation. Its economic sustainability is uncertain and there is also potential for negative social impacts. In addition, UK nuclear development is characterised by a high degree of international interdependency, with firms and public bodies from multiple countries involved in the deployment of new nuclear power stations and with EURATOM (the European Agency for Nuclear Power) playing a central role in these interdependency dynamics.

Major changes in UK energy policy since 2003 have included a revival of political support for new nuclear power. From the 2008 Climate Change Act onwards, there has been steady policy support for new nuclear power, both in terms of investment incentives and regulation. In 2015, the Conservative administration announced a 'policy re-set' for energy and electricity. This re-set was confirmed in the recently published Clean Growth Strategy (CGS), endorsing political commitment to new nuclear power (HM Government, 2017).

At present, the UK has some 9 GW (gigawatts) of operational nuclear capacity, providing about 21% of all UK electricity demand (DUKES, 2017). All but one of these nuclear plants is expected to close before 2030, although no serious plans for closure yet exist. Moreover, unlike anywhere else in Europe, the UK government has ambitions for a significant future build-up of new nuclear capacity. In a 2013 publication (HM Government, 2013), the UK government suggested that new nuclear capacity might be somewhere between 16 GW and 75 GW by 2050, with 16 GW being the minimum target. An increasing reliance on new nuclear is therefore currently planned in the UK as a major contributor to achieving ambitious decarbonisation commitments. It has yet to be determined whether this will be entirely via larger conventional reactors or with a contribution from so-called small modular nuclear reactors (SMRs). While this is a contested issue, some studies consider nuclear power a secure and effective investment for the UK electricity sector (Levi and Pollitt, 2015).

Renewable energy policy

The adoption of renewable resources within the UK's electricity mix has significantly increased over the last ten years, with their overall contribution to UK electricity generation more than tripling between 2008 and 2016 (Figure 5.1), reaching a total share of more than a quarter of the total domestic installed capacity (DUKES, 2017). With favourable natural conditions, strategic investments in research and technological innovation, and public financial incentives, UK renewable electricity generation has been increasingly relying on offshore wind energy, gaining European and international leadership in the sector (Toke, 2011; Higgins and Foley, 2014; Kern *et al.*, 2014). With offshore wind gaining more attention, renewable resources that received significant public support in earlier years (onshore wind and solar) are now facing cuts in public support, while others such as biomass and tidal have problems. For instance, Drax Power Station is controversial for replacing coal with large amounts of imported biomass, and tidal stream technologies are still in development and more research is needed to find optimal locations and reduce costs.

Wind and solar PV clearly possess substantial potential for contributing further to the decarbonisation process (Raugei and Leccisi, 2016). Coupled with strong political support from both major political parties and most of civil society, wind and more recently solar have attracted substantial private investments, assisted by rapid technological innovation and a corresponding reduction in costs. Onshore wind dominated during the period up to around 2010, with offshore wind now the most promising renewable option for the decarbonisation of the UK electricity system, while solar PV has also grown substantially since 2010. Recent official documents clearly emphasise significant political support for expanding offshore wind in the short, medium and long term (HM Government, 2017).

Our research method is described in the *Research process and methods* box. Note that while both qualitative and quantitative methods were employed, we report here only on the qualitative part of the research.

Research process and methods

The mixed method approach used in this research involved qualitative and quantitative methods to identify and assess risks and uncertainties in low-carbon mitigation pathways for the UK electricity system. We used qualitative methods including interviews, surveys, and face-to-face stakeholder engagement to: identify the possible transition pathways for the UK electricity system; identify and assess the risks and uncertainties associated with these identified futures; and develop scenarios. These scenarios were translated to inform the socio-economic model E3ME to test iteratively the techno-economic feasibility of the mitigation pathways on the national level. The qualitative work relied on the co-operation of 95 experts and stakeholders from 33 different organisations, which included government, private organisations, universities, and NGOs (see Table 5.1).

This relatively large group included strong participation from academic participants, who were easier to reach and more able to participate than other stakeholder groups. Nevertheless, there was sufficient diversity among the other stakeholders consulted, including five representatives of private firms (including EDF Energy, currently constructing the nuclear plant at Hinkley Point), three environmental NGOs, and five representatives of government and public agencies.

The engagement process consisted of several stages, starting with a focus group and face-to-case interviews, followed by three stakeholder and expert consultation workshops. The intensive one-on-one interviews with energy sector experts and an expert focus group helped to identify the main drivers of uncertainties in the nuclear sector. It also scoped out categories of the main perceived risks associated

Table 5.1 Summary of stakeholder activities and participants by type of organisation

Stakeholders' engagement activities	Type of organisation and participants				
	Private	Public	Education research	Private public partnership	Total
Focus group: nuclear power (Brighton, 1 July 2016)	–	–	4	–	4
Interviews (London, October–November 2016)	1	2	3	–	6
First stakeholders' consultation workshop: nuclear sector map and scenarios propositions (London, 21 October 2016)	4	1	6	1	12
Second stakeholders' consultation workshop: assessing risks and uncertainties for nuclear expansion (Falmer, 27 March 2017)	7	4	33	1	45
Third stakeholders' consultation workshop: transition mitigation pathways and UK nuclear power (London, 22 September 2017)	7	1	8	–	16
Survey online ('no new nuclear') (July–September 2017)	4	1	6	1	12
Subtotal	23	9	60	3	95

with the new nuclear power scenario on the one hand and the renewable electricity scenario on the other.

The first stakeholders' consultation workshop aimed at co-designing a system map of the UK nuclear sector following the Technology Innovation System structure (i.e. actors/stakeholders, networks, and institutions) using a specific tool created for this purpose (Nikas *et al.*, 2017). This identified the main interactions among the system elements (Cox *et al.*, 2016) and then elicited the preferences and opinions of the stakeholders for the UK's future electricity system mix for 2030 and 2050, bearing in mind the UK commitment to an 80% emission reduction. From this first stakeholders' consultation emerged multiple future electricity mixes, in which nuclear power has a share. However, stakeholders agreed on the broad shape of two scenarios, which are the subject of this chapter. Based on the literature and stakeholder engagement, we developed a list of potential risks and uncertainties associated with the two pathways (Hanger *et al.*, 2016).

In the second stakeholders' consultation workshop, we assessed the identified risks for nuclear power expansion along different dimensions, such as likelihood and severity of impact. For the no new nuclear mitigation pathway, we assessed the identified risks by using an online survey.

Finally, the third stakeholders' consultation workshop was used to discuss the economic feasibility of the two scenarios and to build a complete and detailed narrative of the corresponding mitigation pathways.

Note that due to the need to guarantee stakeholder anonymity it has generally not been possible to attribute particular stakeholder views to specific categories of stakeholder.

Decarbonisation pathways

Based on the critical roles nuclear and renewables may have in the decarbonisation of UK electricity, we introduce two opposing decarbonisation pathways. The first is based on major nuclear expansion and the second is a regime with no new nuclear investment and dominated by renewable resources. These decarbonisation pathways are similar to those known as the 'clockwork' and 'patchwork' scenarios which have been proposed by the Energy Technologies Institute (ETI) (2015).

The 'new nuclear' pathway includes 40 GW of new nuclear capacity. This is highly salient because the UK government relies heavily on new nuclear power in its decarbonisation efforts. Stakeholders at our first workshop, including those with a strong commitment to nuclear power, considered that a level of 75 GW of new capacity by 2050 was not technologically and financially feasible (e.g. delays on building new facilities and the high cost of electricity generation). However, 40 GW of new nuclear capacity might realistically and credibly be achieved according to stakeholders, particularly considering that France had achieved an even higher rate of nuclear construction in the 1970s and 1980s. Despite the inherently controversial and politicised character of decisions on nuclear technology, public opposition to nuclear power (and questions of its legitimacy) have always been muted in the UK.

In the light of sustainable development considerations, the 'no new nuclear' pathway is also clearly salient to any debate about current and future electricity production, and a future without further nuclear capacity is clearly a real possibility. Given radical decarbonisation objectives, such a pathway implies a major expansion in renewable capacity and over the past decade renewables have gained credibility, growing quickly and with ever-reducing costs. High nuclear costs may persuade the financial community that the technology does not deserve support and that there are alternatives in achieving decarbonisation in the absence of further nuclear power. Strong public support for solar PV and offshore wind marks the 'no new nuclear' pathway as highly legitimate, although there is some evidence of public disquiet about onshore wind.

The 'new nuclear' pathway

A 'new nuclear' pathway could be credible given the following circumstance. Politically, a nuclear expansion of 40GW would require a strong and largely bipartisan continuation of the support given to large-scale and centralised technology options in the 2015 policy re-set and the 2017 Clean Growth Strategy, including nuclear power, as part of a continuing commitment to stringent emission reductions by mid-century. This ongoing commitment to nuclear expansion will be strengthened by the view that new nuclear power is vital to energy security, as well as a major contributor to decarbonisation. Geopolitical worries about the UK's need to import higher volumes of gas and oil will support this security argument over the next few years.

The 'new nuclear' pathway foresees a slowdown (and a possible halt) in the reduction of renewable energy costs. At the same time, another barrier to renewables expansion is the difficulty of providing system backup for the intermittent renewable sources that dominate in the UK. The bulk, seasonal electricity storage that this requires continues to be expensive and impractical. Thus for radical decarbonisation, nuclear expansion would be the only proven option. Public opposition to technologies like onshore (and increasingly offshore) wind increases and limits further renewable energy growth. In addition, difficulties remain in implementing major demand-reduction schemes that would avoid some new capacity needs. As expansion in nuclear capacity gains momentum, the possibilities for renewables expansion is in any case constrained. Nuclear power performs well worldwide, both economically and in terms of safety.

The nuclear industry overcomes some of the technological and cost problems currently besetting the European pressurised water reactor (EPR). Future reactor design becomes standardised on the best-performing current technologies, whether of the so-called Generation III+ or the not-yet-commercialised Generation IV, allowing a reduction in costs and subsidies. Global political instability encourages intensive nuclear technology development, both internationally and in the UK, as an apparently secure option, and SMRs could emerge (Department for Business Energy & Industrial Strategy (BEIS), 2017a; Locatelli *et al.*, 2017) as a more flexible option for part of the 40GW capacity, though the feasibility of commercial SMRs is yet to be established. Global instability also reinforces the idea that a strong civil nuclear industry as helpful in maintaining the UK's capacity to possess credible nuclear weapons (Johnstone and Stirling, 2015).

Political support appears to be maintained by recent government decisions and documents, such as the announcement of the follow-up to Hinkley Point C nuclear project (15 September 2016); the 2017 Clean Growth Strategy (HM Government, 2017); and the consultation for a new National Policy Statement (NPS) for future nuclear generation (2026–2035), announced by BEIS in early December 2017 (Department for Business Energy & Industrial Strategy (BEIS), 2017b). Support for research and innovation appears strong, with BEIS's Energy Innovation Programme planning to invest around £180 million in nuclear innovation (HM Government, 2017, p. 50). While a change in UK governing parties may affect such political support, traditional cross-party commitment to nuclear generation persists. Based upon the requirement for similar technologies and skills, interrelationships between civil and military use of nuclear technology further appear to suggest durable political support for nuclear generation.

Yet while steady political support for nuclear research and innovation appears robust, other potential negative consequences may become significant barriers for the implementation of the described pathway. Based on our stakeholder consultations, we identified those that appear to be the most significant among these risks (Third stakeholders' consultation workshop, 2017) with respect their causes/drivers, likelihood, magnitude, and possible effects (see Figure 5.3).

Risks associated with the 'new nuclear' pathway

Conflicting policies and regulations

The nuclear power sector is embedded in a wide-ranging, trans-sectoral institutional framework, which brings significant complexity in the development and application of regulations for new nuclear. The UK nuclear sector is subject to significant national safety and security legislation and to (national and international) laws on the military use of nuclear. This interaction between different policies and regulations can affect nuclear as a source for domestic generation of electricity in several ways. First, the imposition of strict safety and security regulations may lead to increases in the costs of production and maintenance of nuclear plants. Second, potential conflicts with international regulations may lead to long-term international disputes, which could significantly slow down the development of new plants (as these are traditionally developed through international co-operation systems). Brexit may further increase the risk of conflicts between the UK and international legislation on nuclear. Stakeholders considered conflicting policies and regulation, and in particular international legal clashes, as a significant risk both in terms of their likelihood and magnitude, with effects that could in the worst-case scenario lead to a paralysis of UK civil nuclear development.

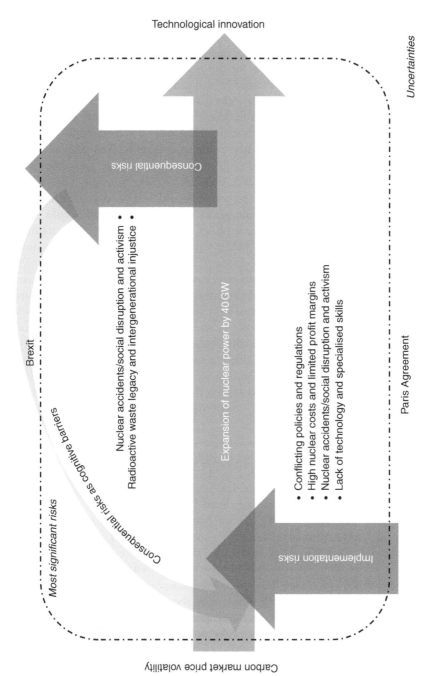

Figure 5.3 Main uncertainties, implementation and consequential risks associated with the 'new nuclear' pathway in the UK.

High nuclear costs and limited profit margins

While it has low operational costs, nuclear electricity suffers from high and often very poorly estimated costs of construction (e.g. Harris *et al.*, 2013). Stakeholders emphasised cost escalation for nuclear reactors and delays in implementation among the main challenges for the sector. High costs for nuclear-based electricity require substantial incentives and subsidies. Such costs may impact low-income consumers, potentially exacerbating energy poverty (Focus group: nuclear power, 2016). Most stakeholders, irrespective of their preferences, regarded this as a high risk, while its impacts are defined as medium-to-marginal if adequately counteracted by ad-hoc public investments. The cost issue has become more acute with financial problems besetting private firms like EDF, Areva, and Toshiba (Thomas, 2017; Vaughan, 2017). However, negotiation between investors and the UK government over a long-term fixed price for electricity well above current wholesale prices could counteract such risks. The CGS addresses the need to reduce costs of nuclear power while maintaining safety by investing in innovation. SMRs, with lower capital requirements, may represent an alternative long-term option to mitigate investment risk, although as yet commercially unproven. SMR development in the UK would require an adequate regulatory framework and siting might prove problematic as SMRs are likely to be most cost-effective if sited near urban areas. Overall, the risk of lack of investment because of high costs is registered as having high likelihood among stakeholders, depending on the evolution of timings and costs for new nuclear power plants, as well as on ongoing uncertainties in electricity prices. In the opinion of stakeholders, the magnitude of the effects of this risk is less certain. Their impact may be limited given the overall political support for new nuclear (see the earlier discussion about the ongoing policy support for the sector).

Nuclear accidents/social disruption and activism

Despite being two separate risks, nuclear accidents and social disruption/activism can be often correlated, as underlined by a majority of our stakeholders. Hence we discuss them as complementary sub-risks of a single category. Managing social acceptability of nuclear expansion per se received very little emphasis among the stakeholders, who underlined that social acceptance of nuclear in the UK is widely diffused, with studies supporting such a perspective (Jay *et al.*, 2014). Yet stakeholders clearly stressed that nuclear accidents elsewhere in the world can strongly affect public perception about nuclear expansion, significantly increasing social opposition. Past experience has shown that even distant accidents have radically shifted countries' positions towards nuclear. This happened for example in Germany, where early nuclear shutdowns were a consequence of the Japanese Fukushima accident (Kepplinger and Lemke, 2016). While the chances of major nuclear accidents and related consequences are regarded as small in general terms, the majority of stakeholders perceive the magnitude of their possible effects on the UK nuclear power as extremely high. Similar accidents can in fact become key triggers in radically increasing activism and social disruption, posing significant threats to the entire UK nuclear sector, especially in any case of geographically close and high-impact accidents (Wheatley, Sovacool *et al.* 2016).

Radioactive waste legacy and intergenerational injustice

There is a high degree of uncertainty regarding the back-end of the nuclear fuel cycle (decommissioning and waste management), especially the failure to date to establish a UK geological repository for high-level wastes. Considering the ethical issues, poor waste management governance, and the scale and longevity of radioactive waste, there is a widespread stakeholder view that the risk has not been fully taken into account in UK nuclear expansion decisions (Second stakeholders' consultation workshop, 2017). A specific issue is the peripheralisation/ marginalisation of communities surrounding nuclear facilities and waste sites. These communities may be prone to manipulation of public opinion by the nuclear industry to develop an apparent social acceptance (Sovacool, 2011, p. 211). The risk of continued failure to develop adequate and democratic waste governance is considered a risk whose magnitude and effects are hard to define precisely. No timely mitigation strategy was advanced by stakeholders.

Lack of technology and specialised skills

Stakeholders have considered this among the most likely and potentially harmful risks to the success of new nuclear power in the UK. While this may seem surprising given the UK's long experience in nuclear technology, the UK is contemplating the commercialisation of at least four overseas nuclear technologies as part of its expansion plan (the EPR, AP1000, ABWR, and Hualong) and possibly more (e.g. SMRs). The greater part of the relevant knowledge about these technologies resides outside the UK, so this significant expansion of novel nuclear technologies and related systems requires development of many new skills for the implementation of such an ambitious and diverse nuclear pathway. The needs for significant technology transfer processes and agreements can be seen as limits for a domestic implementation of the sector. In addition, Brexit, which may lead to a slowdown in technological relationships with traditional UK nuclear partners such as France, may further increase both the likelihood and magnitude of such risk. Finally, the lack of people with specialised skills in nuclear regulation is considered as a particular risk for the sector by the stakeholders. Lack of capacity among regulators could lead to a significant misalignment between the ambitions for the UK nuclear sector and the means to fulfil them (First stakeholders' consultation workshop, 2016).

The 'no new nuclear' pathway

In this narrative, much depends on reasons for the failure of the current policy of major nuclear expansion. In a 'no new nuclear' pathway, raising finance for new nuclear power projects becomes impossible. It took EDF several years to assemble the £25 billion financing cost for Hinkley Point C and the successful construction of this first new project is far from certain, especially given the fragility of the finances of EDF and problems in quality control in the manufacturing of critical components in France. In this narrative, the government is unwilling to allow future nuclear projects to risk costing consumers the equivalent of the £30 billion excess costs that Hinkley involves (National Audit Office, 2017). At the same time the world market for new nuclear stations fails to pick up, putting further pressure on nuclear vendors, several of whom (Toshiba/Westinghouse and EDF/Areva) are already facing some major financial problems.

The recent cost reductions in renewable technologies would continue and make renewables highly competitive in market conditions. The offshore wind sector, offering opportunities for a significant number of large-scale sites, develops strongly and offers increasingly competitive costs. Solar PV is also widely deployed, both in the form of solar farms and in building design. Currently uneconomic renewables, such as tidal power, gradually become more cost-competitive and local biomass use becomes more popular. Problems of both seasonal and short-term bulk electricity storage are overcome at reasonable prices, allowing renewables to achieve a large majority share of electricity generation despite the fact that they are not dispatchable. More comprehensive interconnections with other countries would help. However, energy security might constitute a high risk and uncertainty unless such storage systems can be developed and work with high reliability. While the development of CCS is imaginable in this pathway (as also the case for the 'new nuclear' pathway), this is not considered here as the commercial development and deployment of CCS is far from certain.

Such a renewables-only expansion pathway in the UK electricity system involves a significant change of direction and risks (see Figure 5.4). This means that a 'no new nuclear' pathway would be accompanied by a major new priority for demand-reducing measures, allowing an 80% or more emission reduction commitment to be met with a lower total installed capacity. Some renewables in the UK enjoy policy support similar to that surrounding new nuclear. Both traditional and recent policies in the UK seem to secure robust short- and medium-term support for at least some forms of renewable energy. Scope for reductions in energy demand across all consumption sectors, especially in households, is also recognised and supported by policy, hence further facilitating the deployment of renewables. Wind energy, and particularly offshore wind technologies, can count on strong support from British policymakers, identifying it as a key strategic technology for the country. For instance, the 2017 Clean Growth Strategy (CGS) commits a total of more than £700 million for offshore

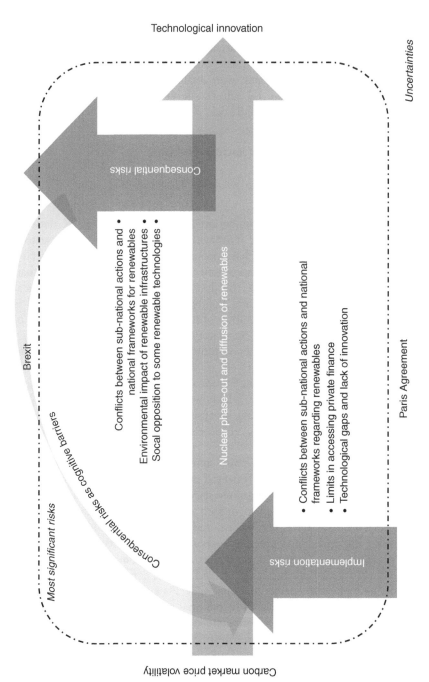

Technological innovation

Uncertainties

Consequential risks

• Conflicts between sub-national actions and national frameworks for renewables
• Environmental impact of renewable infrastructures
• Socal opposition to some renewable technologies

Brexit

Nuclear phase-out and diffusion of renewables

Consequential risks as cognitive barriers

Most significant risks

Paris Agreement

Implementation risks

• Conflicts between sub-national actions and national frameworks regarding renewables
• Limits in accessing private finance
• Technological gaps and lack of innovation

Carbon market price volatility

Figure 5.4 Main uncertainties, implementation and consequential risks associated with the 'no new nuclear' pathway in the UK.

wind, with more than £500 million for contracts with generating firms (in the form of auctions) and additionally £150 million to be invested in offshore technological innovation. In this pathway, there would also be strong support for solar PV, as well as further R&D and potentially deployment of other currently non-commercial renewables, especially tidal power and possibly wave energy. The current policy of discouraging and even disallowing onshore wind would be reversed. Overall, doubts about the effective level of interest in renewables by current UK policymakers (Gillard and Lock, 2017, p. 646) would be reversed.

Risks associated with the 'no new nuclear' pathway

Conflicts between sub-national actions and national frameworks regarding renewables

Conflicting conditions between sub-national initiatives and the UK national framework for renewables have systematically characterised the sector's dynamics. For instance, Scotland has adopted a wide array of interconnected policies favouring renewables deployment to a much greater extent than England and Wales, facilitated by the devolution (ceding) of some political authority to Scotland (McEwen and Bomberg, 2014). However, Scotland is facing limits to the full implementation of its renewables potential because of conflicts between Scottish and UK regulatory frameworks (Cowell *et al.*, 2017). A typical example of such conflict is provided by tidal energy. Seen as an exploitable resource in Scotland (Neill *et al.*, 2017), tidal attracts very little attention from policymakers and institutional investors in England (Lamy and Azevedo, 2018). As a consequence, tidal power keeps on, overall, being marginalised, including in Scotland. This is because, constrained by the unfavourable regime surrounding tidal at UK level, Scotland has had to limit strategies to small-scale piloting experiments.

Community-led initiatives have also suffered similar limits over the years. UK renewables development has benefited from many small-scale community-led initiatives. However, diverging interests, objectives, and scope have shown the limits of a synergistic implementation of community-led initiatives and national electricity systems, regulations, and policies, and recent UK government policy has withdrawn previous support for community energy projects. While generally providing social, economic, and environmental benefits at the local level (Bracken, Bulkeley, and Maynard, 2014), community-led initiatives have a limited effect on a larger, national scale. As a consequence, community-led initiatives tend to be neglected by national policymakers and institutional investors (Saintier, 2017) and government subsidies and tax incentives for community energy projects were withdrawn in 2015 (Community Energy England, 2018). Stakeholders defined this set of risks as already affecting the implementation of renewables. Efforts to integrate local, regional, and national frameworks, as well as increasing policymaker awareness of community-led initiatives, were suggested as the main solutions for mitigating the risk.

Limits in accessing private finance

Access to private finance for renewables in the UK is determined by a multiplicity of factors, leading to relatively high degrees of risk and uneven expectations. More specifically, renewables' access to private finance is affected by: (a) costs of the adopted technologies; (b) the general volatility of market prices for electricity; (c) variations in prices of other resources (i.e. fossil fuels and nuclear); and (d) uneven and/or unexpected changes in carbon markets. Technology costs are in many ways the main driver in determining private financial support for renewables, with technologically mature resources such as wind and solar more favourably supported than niche ones. More generally, increasing interest from private institutional investors in renewables in the UK, driven by the rapid pace of cost reductions, makes the likelihood of a lack of private finance appear relatively limited (Winch and Wynn 2017, pp. 22–23). Nevertheless, the majority of stakeholders underlined how the presence of many types of risk concerning private finance means that the overall risk cannot be regarded as unlikely. In addition, substantial reductions in private finance would lead, in the opinion of several stakeholders, to significant damage for the technologies/resources affected. Stakeholders unanimously identified better co-ordination between national policies and private finance as the best way to mitigate the risk.

Environmental impact of renewable infrastructures and social opposition to some renewable technologies

As already suggested for the 'new nuclear' pathway, two complementary risks are analysed here under a single perspective. As addressed by those stakeholders who raised these two risks, these are often interlinked. While nuclear is generally perceived as more threatening to nature and environmental preservation, questions have also been raised on the possible environmental impacts of renewable technologies, with special regard to those that imply large-scale development of new infrastructure in previously untouched natural areas. Apart from unwelcome visual impacts, wind deployment using ever-larger blades has raised concerns for the possible effects on migratory birds (Hattam, Hooper, and Papathanasopoulou, 2017; Hooper, Beaumont, and Hattam, 2017). Stakeholders suggested that, despite British public opinion being generally favourable to renewable technologies (Allen and Chatterton, 2013), these risks have contributed to the rise of social opposition to some renewable technologies. In addition to environment-related social opposition, some stakeholders suggested that increases in electricity tariffs due to the need for financial support for new clean technologies could lead to opposition. Further objection to visually intrusive renewable solutions is likely, such as with large-scale solar farms and onshore wind power plants (and already, for example in mid-Wales, new onshore wind developments are politically virtually impossible). Stakeholders defined this set of risks as likely to happen and in part already happening. At the same time the small size and localised nature of social opposition to renewables makes this a limited risk in terms of its overall effects on the renewables sector.

Technological gaps and lack of innovation

UK leadership in offshore wind technologies is becoming more visible (Colmenar-Santos *et al.*, 2016). However, at the same time the UK lacks expertise and technological ownership with respect to most other renewables. As a consequence, such renewables' implementation continues to rely on technology imports, leaving a possible energy security issue. While stakeholders agreed in affirming that an adequate policy to favour deployment of the technologies and skills may contribute to reduce the likelihood of this risk, they also affirmed how, in the case of inaction, the effects may be significant. Several stakeholders affirmed that inaction would lead to a significant loss of competitiveness for the UK in the renewables sector, affecting UK prospects in terms of sector-based economic growth. The issue of ensuring that there is cost-effective bulk storage for intermittent renewables at high levels of system penetration remains a serious risk, even though several different strategies (including demand response) are available (Grünewald *et al.*, 2011). Better interconnectors would help but they are themselves dependent on a surplus availability of power in other countries. As in the case of 'new nuclear', stakeholders identified the lack of skills of sector regulators as a key risk for renewables implementation.

Uncertainties associated with the two pathways

While risks are defined in relatively specific ways as potential threats to decarbonisation pathways, uncertainties are more general but potentially relevant unknowns, the effects of which could either be positive or negative. Four main types of uncertainty emerge, three of them typically affecting decarbonisation pathways anywhere in the world and one specifically affecting the UK. We identified two traditional drivers (carbon market price volatility and technological innovation) and two recent drivers (the Paris Agreement on climate change and Brexit), all four affecting both pathways. Except for the Paris Agreement, generally seen by stakeholders as a driver for reducing uncertainty, the other drivers were seen as sources of significant uncertainty for UK electricity generation. This section describes how each of these drivers leads to uncertainty for both the 'new nuclear' and the 'no new nuclear' pathways.

With a long history of uneven fluctuations, electricity prices represent a significant source of uncertainty for investors and other stakeholders engaged in the implementation of both new nuclear and renewable electricity. Price fluctuations, or expectations of fluctuations, affect long-term investments in the sector. Yet rapid variations in electricity prices linked to shifts in competing resource prices (i.e. for fossil fuels) may lead to significant uncertainty in the medium and short term. Public intervention and long-term contracts with fixed electricity prices can compensate for such uncertainties but require significant state engagement and action. Stakeholders involved in the assessment almost unanimously agreed that this is a major driver of uncertainty, in particular for the nuclear sector.

Rapid advances in energy-related technological innovation may lead to significant changes in the electricity framework, in the UK as well as in the rest of the world. The breakthrough in new, economically, socially, and environmentally competitive technologies may represent a key element in determining changes within the electricity mix and may lead to the replacement of obsolete incumbent technologies. While possibly appearing as a risk, uneven technological innovation has been included as a driver of uncertainty due to its unpredictability. While it is possible that new technologies breakthrough in both new nuclear and non-nuclear sectors, it is impossible to forecast these changes. For instance, nuclear fusion represents a long-lasting chimera for nuclear and electricity generally, though its effective realisation could lead to a radical change in the electricity mix worldwide. Technology commitments at national level serve to mitigate the uncertainties concerning technological innovation. When analysing the CGS sections on investment in low-carbon research and innovation (HM Government, 2017, pp. 100–101), it can be seen that the UK is significantly committed to stimulating technological innovation. In detail, the CGS promises significant investments for relevant research and innovation (more than £900 million to 2021), devoting more than 50% of these investments to nuclear power while limiting investment in renewables research to just 20% of the total.

The December 2015 Paris Agreement on Climate Change contains highly ambitious goals in terms of mitigation of carbon dioxide (CO_2) emissions and therefore has potentially significant impacts for the energy sectors for all signatory countries. However, the agreement still lacks full design of its regulatory framework, leaving significant uncertainty margins that affect electricity stakeholders in the UK and elsewhere. Stakeholder opinions in our research suggest that the Paris Agreement may act as a driver in reducing uncertainty for both the 'new nuclear' and 'no new nuclear' pathways, thanks to its quantified pledges to be delivered at national and international levels. The great majority of stakeholders suggested that, because both pathways have low-carbon ambitions, the Paris Agreement could reduce uncertainties and ultimately favour the implementation of either pathway. A limited number of stakeholders judged the Paris Agreement as possibly neutral in driving the selected pathways, while no stakeholder identified the agreement as a driver of additional uncertainty.

The referendum binding the UK to leaving the European Union (Brexit) was widely recognised as a significant driver of uncertainty for the implementation of both the 'new nuclear' and 'no new nuclear' pathways. Stakeholders stressed that the UK exit from EURATOM creates the biggest Brexit-related uncertainties for UK new nuclear. Uncertainty about possible limits for partnerships with European firms has also been recognised as potentially threatening for the nuclear sector. On the other hand, some stakeholders have suggested possible bureaucratic burden reductions and easing of public funding as positive Brexit effects for nuclear power. Brexit was also recognised by the great majority of stakeholders as a major driver of uncertainty for renewable electricity production in the UK. Brexit-related uncertainties are most importantly related to investment, doubts about the robustness of the UK's mitigation commitments, exit from the European Emissions Trading System (EU ETS), and more difficult relationships with European partners.

Conclusions

The choice of focusing on nuclear power for the UK case study was based on the strong commitment of the UK government to new nuclear as a major component of its climate mitigation strategy, when no other EU member state has made commitments greater than marginal further nuclear construction. If the UK government is successful in pursuing this strategy there could, however, be significant implications for mitigation policies in many other countries. The two pathways investigated – involving, respectively, a very high rate of nuclear construction and an abandonment of nuclear power – were deliberately framed as possible 'extremes' in relation to the future of mitigation in the UK, though both have credible storylines.

Our results have limitations, partly because our stakeholder sample was skewed towards academics with relatively limited consultation of other stakeholders. The need for strict stakeholder anonymity (especially stressed by industry participants) also means that we cannot report attributions of particular stakeholder views to stakeholder categories. Nevertheless, it is striking that in relation to assessment of several of the risks considered, there was a high degree of consensus about the existence and seriousness of risks across stakeholders with widely differing interests and views. A good example is the related risks of high cost and problematic financing that could affect both a high nuclear and a no nuclear pathway – a serious risk expressed by a large majority of stakeholders. Stakeholders also often showed substantial agreement on the most effective actions that could mitigate several of the risks. These included greater public-sector commitment to technology and skill development and the need to avoid conflict between different spheres of policy and regulation at sub-national, national, and international levels.

Both pathways were shown to have significant risks attached, often common to both pathways. However, it was clear that stakeholder views suggested that the 40 GW nuclear pathway had particularly difficult risks attached to it, involving large and controversial state and consumer funding. Recent developments, including very high nuclear costs, slow progress in nuclear construction, and regulatory problems associated with Brexit tend to give external confirmation of the severity of risks to a high nuclear future. However, UK government sees the need to support new nuclear as an option in case no better and credible technological alternatives emerge (e.g. low-cost tidal and biomass and full development and deployment of CCS) to reduce risk and uncertainty for energy security if heat, transportation, and industry would be electrified. Moreover, the combination of renewable generation with substantial energy demand reductions will require more policy attention. While this suggests that more weight may need to be placed on renewables in the UK's low-carbon future, there remain significant risks here as well. The most serious among them is the development of cheap enough bulk seasonal storage to overcome intermittency problems and the high costs that may still attach to developing many more GW of renewable capacity.

Our final conclusion is therefore that both a mainly new nuclear and a mainly renewable pathway need significant work to overcome a range of important risks. Research here needs to engage with a wider range of stakeholders and concentrate on ways in which public support and legitimacy can be secured for mitigation strategies that will probably involve significant future public or consumer subsidy.

References

Allen, P., Chatterton, T. (2013). Carbon reduction scenarios for 2050: An explorative analysis of public preferences. *Energy Policy* 63, 796–808.

Bracken, L.J., Bulkeley, H.A., Maynard, C.M. (2014). Micro-hydro power in the UK: The role of communities in an emerging energy resource. *Energy Policy* 68, 92–101.

Colmenar-Santos, A., Perera-Perez, J., Borge-Diez, D., dePalacio-Rodríguez, C. (2016). Offshore wind energy: A review of the current status, challenges and future development in Spain. *Renewable and Sustainable Energy Reviews* 64, 1–18.

Community Energy England (2018). *Community Energy State of the Sector 2018.* Available at: https://communityenergyengland.org/files/document/169/1530262460_CEE_StateoftheSectorReportv.1.51.pdf [accessed 24 August 2018].

Cowell, R., Ellis, G., Sherry-Brennan, F., Strachan, P.A., Toke, D. (2017). Energy transitions, sub-national government and regime flexibility: How has devolution in the United Kingdom affected renewable energy development? *Energy Research & Social Science* 23, 169–181.

Cox, E., Lieu, J., MacKerron, G., Alvarez-Tinoco, R. (2016). *Transitions Pathways and Risks Analysis for Climate Change Mitigation and Adaptation Strategies. D3.2 Context UK Nuclear Power.* TRANSrisk Project. EC Horizon 2020: ID 642260. Available at: http://transrisk-project.eu/sites/default/files/Documents/D3_2_CaseStudy_UK.pdf [accessed 24 February 2017].

Department for Business Energy & Industrial Strategy (BEIS) (2017a). *Small Modular Reactors competition: phase one. A competition to gather evidence to inform policy development on Small Modular Reactors (SMRs).* Available at: www.gov.uk/government/publications/small-modular-reactors-competition-phase-one [accessed 12 April 2018].

Department for Business Energy & Industrial Strategy (BEIS) (2017b). *Consultation on the sitting criteria and process for a new National Policy Statement for nuclear power with single reactor capacity over 1 Gigawatt beyond 2025.* Available at: www.gov.uk/government/uploads/system/uploads/attachment_data/file/665590/061217_FINAL_NPS_Siting_Consultation_Document.pdf [accessed on 9 December 2017].

DUKES (2017). *Digest of United Kingdom Statistics, July 2017. Department for Business, Energy & Industrial Strategy.* Available at: https://assets.publishing.service.gov.uk/government/uploads/system/uploads/attachment_data/file/643414/DUKES_2017.pdf [accessed 30 April 30 2018].

Energy Technologies Institute (ETI) (2015). *Options Choices Actions-UK scenarios for a low carbon energy system*. Available at: www.eti.co.uk/insights/options-choices-actions-uk-scenarios-for-a-low-carbon-energy-system/ [accessed 13 November 2017].

First stakeholders' consultation workshop (2016, 21 October 2016). Transitions Pathways and Risks Analysis for Climate Change Mitigation and Adaptation Strategies. TRANSrisk project. EC Horizon 2020: ID 642260. London.

Focus group: nuclear power (2016, 1 July). Transitions pathways and risk analysis for climate change mitigation and adaptation strategies. TRANSrisk Project. EC Horizon 2020: ID 642260. SPRU-University of Sussex, Falmer.

Gillard, R., Lock, K. (2017). Blowing policy bubbles: rethinking emissions targets and low-carbon energy policies in the UK. *Journal of Environmental Policy & Planning* 19(6), 638–653.

Grünewald, P., Cockerill, T., Contestabile, M., Pearson, P. (2011). The role of large scale storage in a GB low carbon energy future: Issues and policy challenges. *Energy Policy* 39(9), 4807–4815.

Hanger, S., van Vliet, O., Bachner, G., Kontchristopoulos, Y., Lieu, J., Alvarez-Tinoco, R., Carlsen, H., Suljada, T. (2016). *Transitions Pathways and Risks Analysis for Climate Change Mitigation and Adaptation Strategies. D5.1 Review of key uncertainties and risks for climate policy*. TRANSrisk Project. EC Horizon 2020: ID 642260. Available at: http://transrisk-project.eu/sites/default/files/Documents/D5.1%20Review%20of%20key%20uncertainties%20and%20risks%20for%20climate%20policy.pdf [accessed 13 December 2017].

Harris, G., Heptonstall, P., Gross, R., Handley, D. (2013). Cost estimates for nuclear power in the UK. *Energy Policy* 62, 431–442.

Hattam, C., Hooper, T., Papathanasopoulou, E. (2017). A well-being framework for impact evaluation: The case of the UK offshore wind industry. *Marine Policy* 78, 122–131.

Higgins, P., Foley, A. (2014). The evolution of offshore wind power in the United Kingdom. *Renewable and Sustainable Energy Reviews* 37, 599–612.

HM Government (2008). *Climate Change Act 2008: Chapter 27*. The Stationery Office, UK.

HM Government (2013). *The UK's Nuclear Future*. Available at: https://assets.publishing.service.gov.uk/government/uploads/system/uploads/attachment_data/file/168048/bis-13-627-nuclear-industrial-strategy-the-uks-nuclear-future.pdf [accessed 24 August 2018].

HM Government (2017). *Clean Growth Strategy: Leading the way to a low carbon future*. Available at: www.gov.uk/government/uploads/system/uploads/attachment_data/file/651916/BEIS_The_Clean_Growth_online_12.10.17.pdf [accessed 18 October 2017].

Hooper, T., Beaumont, N., Hattam, C. (2017). The implications of energy systems for ecosystem services: A detailed case study of offshore wind. *Renewable and Sustainable Energy Reviews* 70, 230–241.

Jay, B.M., Howard, D., Hughes, N., Withaker, J., Anandarajah, G. (2014). Modelling Socio-Environmental Sensitivities: How Public Responses to Low Carbon Energy Technologies Could Shape the UK Energy System. *The Scientific World Journal* Volume 2014, Article ID 605196.

Johnstone, P., Stirling, A. (2015). Comparing Nuclear Power Trajectories in Germany and the UK: From 'Regimes' to 'Democracies' in Sociotechnical Transitions and Discontinuities. *SPRU Working Paper Series (SWPS)* 2015–18, 1–86.

Kepplinger, H.M., Lemke, R. (2016). Instrumentalizing Fukushima: Comparing Media Coverage of Fukushima in Germany, France, the United Kingdom, and Switzerland. *Political Communication* 33(3), 351–373.

Kern, F., Smith, A., Shaw, C., Raven, R., Verhees, B. (2014). From laggard to leader: Explaining offshore wind developments in the UK. *Energy Policy* 69, 635–646.

Lamy, J.V., Azevedo, I.L. (2018). Do tidal stream energy projects offer more value than offshore wind farms? A case study in the United Kingdom. *Energy Policy* 113, 28–40.

Levi, P.G., Pollitt, M.G. (2015). Cost trajectories of low carbon electricity generation technologies in the UK: A study of cost uncertainty. *Energy Policy* 87, 48–59.

Locatelli, G., Fiordaliso, A., Boarin, S., Ricotti, M.E. (2017). Cogeneration: An option to facilitate load following in Small Modular Reactors. *Progress in Nuclear Energy* 97(Supplement C), 153–161.

McEwen, N., Bomberg, E. (2014). Sub-state Climate Pioneers: The Case of Scotland. *Regional & Federal Studies* 24(1), 63–85.

National Audit Office (2017). *Hinkley Point C.* Available at: www.nao.org.uk/wp-content/uploads/2017/06/Hinkley-Point-C.pdf [accessed 24 April 2018].

Neill, S.P., Vögler, A., Goward-Brown, A.J., Baston, S., Lewis, M.J., Gillibrand, P.A., Waldman, S., Woolf, D.K. (2017). The wave and tidal resource of Scotland. *Renewable Energy* 114, 3–17.

Nikas, A., Haris, D., Lieu, J., Alvarez-Tinoco, R., Charisopoulos, V., van der Gaast, W. (2017). Managing stakeholder knowledge for the evaluation of innovation systems in the face of climate change. *Journal of Knowledge Management* 21(5), 1013–1034.

Raugei, M., Leccisi, E. (2016). A comprehensive assessment of the energy performance of the full range of electricity generation technologies deployed in the United Kingdom. *Energy Policy* 90, 46–59.

Saintier, S. (2017). Community Energy Companies in the UK: A Potential Model for Sustainable Development in 'Local' Energy? *Sustainability* 9. doi:10.3390/su9081325.

Second stakeholders' consultation workshop (2017). Transitions Pathways and Risks Analysis for Climate Change Mitigation and Adaptation Strategies. Assessing Risks and Uncertainties for nuclear expansion. TRANSrisk Project. EC Horizon 2020: ID642260. *ESRC Nuclear Futures: Re-making sociotechnical research agendas Seminar 4.* Bramber House Conference Centre, Brighton, 27–28 March, University of Sussex.

Sovacool, B.K. (2011). *Contesting the future of nuclear power. A critical global assessment of atomic energy.* World Scientific, Singapore.

Third stakeholders' consultation workshop (2017, 22 September). *Transitions Pathways and Risks Analysis for Climate Change Mitigation and Adaptation Strategies. Transitions Mitigation Pathways and UK Nuclear Power.* TRANSrisk Project. EC Horizon 2020: ID 642260. London.

Thomas, S. (2017). China's nuclear export drive: Trojan Horse or Marshall Plan? *Energy Policy* 101, 683–691.

Toke, D. (2011). Special issue on offshore wind power. *Energy Policy* 39(2), 495.

Vaughan, A. (2017) 'Westinghouse bankruptcy move casts shadow over world nuclear industry.' *Guardian.*

Wheatley, S., Sovacool, B.K., Sornette, D. (2016). Of disaster and Dragon Kings: A statistical analysis of nuclear power incidents and accidents. *Risk Analysis* 37(1), 1–17.

Winch, H., Wynn, G. (2017). *The renewable energy infrastructure investment opportunity for UK pension funds.* Available at: http://greenfinanceinitiative.org/wp-content/uploads/2017/11/Final-Report-14.11.2017.pdf [accessed 30 April 2018].

Part III

Pathways towards renewable electricity systems

6 Chile

Promoting renewable energy and the risk of energy poverty in Chile

Luis E. Gonzales Carrasco and Rodrigo Cerda[1]

Introduction

In the last six years, one of the most important global issues has been the challenge of climate change and its economic impacts. One of the main determinants of the acceleration of climate change is emissions of greenhouse gases (GHG). According to the International Energy Agency (IEA, 2018), during the three-year period 2014–2016, global emissions levelled off at around 32.1 million tons of CO_2. Of this total, 42% came from electric generation, 23% from transportation (mostly land transport at 17%), 19% from manufacturing and construction, with the rest attributed to other sectors. This deceleration in worldwide emissions growth is mainly due to the substitution of fossil fuels for renewable energies.

Among the renewable energy technologies, solar energy has made one of the biggest market impacts in recent years. In Latin America, for 2016, there is installed solar capacity of 2828 MW, equivalent to 4% of the installed capacity of the United States, 2% of China, and 1.5% of Europe. According to the International Renewable Energy Agency (IRENA, n.d.), Chile is leading the region in its exploitation of solar potential, with 56.7% of Latin America's installed capacity during 2011–2016. Building on this positive trend, and due to the continued international pressure to mitigate the emissions of GHGs, a carbon tax was implemented to complement the increase in clean energy with an incentive to reduce emissions elsewhere. Nevertheless, this push towards substituting fossil fuels with renewable energy also implies an expected rise in electricity prices in the medium term. The extent depends on the electricity mix but may negatively impact the cost of energy for households, disproportionally affecting the poor because of their low income disposal to cover additional cost in their electric bills after the implementation of the tax.

In particular, as an emerging economy Chile faces the challenge of continuing its development path while at the same time maintaining economic growth, social welfare, and environmental action. Poverty in Chile has declined significantly since the early 1990s. This is measured in two ways: first, using the traditional income measure, and second, by the new methodology of multi-dimensional poverty (Ministerio de Desarrollo Social, 2017). In traditional measurement, the poverty line is basically determined by the cost of a food basket. The new methodology of multidimensional poverty includes other factors that relate to vulnerability, such as access to education, health, work and social security, housing and networks, and social cohesion.

In either approach, access to energy and its costs seem to be absent. There are several reasons why it is important to include them as a relevant factor. The first is that climate change will produce (and is already producing) impacts of various kinds, including increases in average temperatures, droughts, and extreme weather events (IPCC, 2014). These phenomena are associated with increases in demand for energy to keep our homes and workplaces at an adequate temperature. In this way, it is expected that energy demand will grow to cope with climate change. A second reason is that to mitigate climate change it is necessary to substitute fossil-fuel-intensive technologies for those with lower emissions of GHGs. To achieve this, policies are being used that may raise energy prices; for example, if a carbon dioxide tax is imposed it particularly affects the electricity sector. This second reason suggests an increase in for households of expenditure on energy, therefore energy expenditure can become a significant part of the family budget, especially for those most vulnerable. This would imply that any increases in prices could place pressure on family budgets, adversely influencing the socio-economic wellbeing of those affected. Thus, in line with this book's risk framing, energy poverty is a consequential risk resulting from Chile's ongoing low-carbon transition.

The variations in climatic regions and resource potential also need to be considered in household energy consumption. In Chile, there are three thermal zones. The first zone is a 'hot' zone in the north which mainly comprises the Atacama Desert and Valleys, where solar power development is mainly concentrated with temperatures reaching up to 40°C degrees. The second thermal zone is in central Chile with a 'temperate climate' between 8–35°C degrees. The main metropolitan areas are situated in this thermal zone; they are the wealthiest part of the country and make up around 60% of Chile's population. The third zone is in the Patagonia region in the southern Andean Mountains where there is high potential for hydro power. This area is sparsely inhabited and income levels are lower.

Overall, electricity prices in Chile high because the electricity mix is imported, including natural gas. There is potential for hydro power but with reservations due to climate change, and coal is not popular due to social pressure which inhibits further development.

In this climatic and socio-economic context, this chapter analyses a low-carbon pathway as manifested around the key idea of incentivising renewable energy through a carbon tax, as introduced in the electricity-generation sector in Chile in January 2017. The particular focus is on the potential energy poverty risk in the absence of a compensation policy, with the aim of deconstructing this potential negative consequence by measuring energy poverty – also known in the literature as 'fuel poverty' – and the degree of energy vulnerability of Chilean households; more concretely, this is done by determining how many households could fall into energy poverty and which specific households would be affected if energy prices were to increase. This tax, by increasing the prices of the electricity sector, indeed decreases poor households' budgets. Thus, focused compensation measures for those most negatively impacted by rising electricity prices could ensure some distributive justice and mitigate this risk. Such compensation measures have limited fiscal costs, which makes them quite attractive. Additionally, non-quantifiable risks associated with a low-carbon energy transition are based on expert perception only and stakeholder perspectives are considered to make a robust analysis in this chapter.

The current Chilean low-carbon pathway

The geographic characteristics of Chile make the country one of the most potential markets for renewable energy expansions. Specifically, the leadership of Chile in solar investments is observed through the significant and fast growth of installed capacity from 2012 to 2017. According to the International Renewable Energy Agency (IRENA, n.d.), in 2012 Chile had 2 MW of solar photovoltaic (PV) installed, and by 2017 was leading in Latin America with 2110 MW of solar PV installed. This 'solar revolution' means an average rate of 302% growth by year. Comparing with the rest of the region, Chile has 58.6% of the installed capacity in 2017.

Four main drivers explain this expansion. First, the observed average growth of the per capita electricity demand of around 4% in the last 46 years. Second, the natural potential of Chile concentrated in the Atacama Desert in the north of Chile, the driest desert in the world with an estimated solar potential for electricity generation of nearly 1000 GW or five times the present peak load of all South America. (Jiménez-Estevez *et al.*, 2015). The third driver is the absence and cost of possible substitutes like the natural gas that in the past achieved US$15 per thousand cubic feet. In the case of coal there is an official agreement for decarbonisation, signed by the industry and the government, that prevents new investment in this fuel in the future (Generadores de Chile, 2018), plus carbon pricing through the implementation of carbon tax. The fourth driver is that the success of this industry was based without fiscal transfers or subsidies. An example of this is the tender for the sale of electric power in 2016, where the company Solarparck Corp Tecnologica was awarded 120 MW of solar energy at US$29.10 MW per hour, practically half the cost of a coal-fired plant in Chile.

To take advantage of these four drivers, actions are needed to: (1) create a more sustainable electricity sector by 2025 to facilitate the entrance of new competitors; (2) interconnect the two main electricity systems, the Central Interconnected System (Sistema Interconectado Central, SIC) and the Northern Interconnected System (Sistema Interconectado del Norte Grande, SING); and (3) promote energy efficiency and clean technologies.

The implementation of these actions requires, however, significant private investment in the sector, which is expected to trigger economic growth.

Recently, the government invited tenders to provide 12.430GWh per year, starting in 2021, in the SIC and SING systems. This call constituted an important step forward in the implementation of the energy agenda. A total of 84 companies participated in the tender process and the electricity supply was awarded to companies that offered to sell electric power to the market at US$47.6 per MWh. This price is much lower than those currently set in the market. Importantly, two-thirds of the investment will be in solar and wind generation (Ministerio de Energía de Chile, 2016).

In addition to the Chilean economic efforts carried out in the last 40 years, the country has been engaged in efforts to mitigate the negative effects of climate change. One of the important efforts was the approval of a law to provide incentives promoting non-conventional energy sources (NCES). The law states that by 2025, 20% of total energy would correspond to NCES (Ley No 20257, 2008).

The current public policy agenda on energy stipulates that 45% of the increase in electricity supply in the 2014 to 2025 period should come from NCES. Chile is also including new policy instruments to mitigate GHG emissions. In the 2014 tax reform, the government introduced a tax on CO_2 emissions and local air pollutants (SO_x, NO_x, and particulate matter) from fixed specific sources. The tax on CO_2 emissions was set at US$5 per ton. In addition, a car tax was imposed based on NO_x car emissions. Nevertheless, an expected correction to this tax is required.

According to the International Energy Agency (IEA, 2018), in 2012 the average global CO_2 emissions per capita were 4.5 tons per year, with Chile having a similar figure. Those numbers are much lower than the Organisation for Economic Co-operation and Development (OECD) average of 97.2 tons per person. In the Latin America region, Chile accounted for 4.7% of total emissions.

Chile, as a member of the UN's Framework Convention on Climate Change (UNFCCC), has committed to reducing GHG emissions by 30% by 2030 (Gobierno de Chile, 2015). The mitigation efforts should occur in different sectors that produce GHGs, focusing on: energy, industry, mining, and other sectors using fossil fuels; processes in the industrial sector; use of land; and waste. To achieve the reduction in GHGs by 2030, Chile committed to reducing its GHGs emissions by 35% to 45% vis-à-vis its 2007 levels, provided it is possible to keep the pace of economic growth and obtain international grants to finance the additional required measures to attain the objective. In the land management sector, Chile committed to the restoration of 100,000 hectares of forestry, which corresponds to the reduction of 600,000 tons of GHG emissions per year.

To tackle this challenge of a low-carbon pathway, policy measures have been taken and implemented. One of the most representative for the market was the carbon pricing signal. After a broader discussion in academia, congress, and civil society, Chile decided to implement a carbon dioxide tax starting at $5/ton of carbon dioxide equivalent as its carbon pricing strategy. This measure was approved in 2014 and implemented recently in 2017, collecting its revenues at the moment less than quinquennia.

At this point, and with the best information, simulations of possible scenarios of both different levels of carbon tax rates and electricity generation mixes have been conducted by different projects like the Mitigation Action Plans and Scenarios for Chile (MAPS Chile, 2014) where macroeconomic impacts are observed with the interaction of electric generation models that represents the expansion plans in the electricity sector and provide the resulting prices of technologies mixes for industry and households.

Using this context of Chile's low-carbon emission pathway and the aggregate information of macroeconomic and energy simulations, four scenarios were developed where energy poverty could emerge as a consequential risk resulting from the promotion of renewable energy by means of a carbon dioxide tax.

Four scenarios to understand energy poverty

To understand possible negative consequences – in this case energy poverty – of continuing this pathway, scenarios help address uncertainties about the future role of renewables in the Chilean energy sector. Across these scenarios any low-carbon pathway needs to support a continuous growth rate appropriate for an emerging country, where the rate of economic growth in capita terms is around 3.5% (Benavente, Gonzales, and Díaz, 2014). This pathway includes the objectives of an electricity generation matrix with a low presence of coal and other fossil-fuel sources, plus the incorporation of renewable technologies like solar and wind. At the same time, this pathway shall reduce negative externalities produced by pollution from activities such as mining and manufacturing. Thus in the near future Chile sees itself as a developed economy that bases its production on environmentally friendly technologies, generating opportunities for them to be included in the electric generation mix of its population in many ways.

The first scenario consists of moderate solar development. The main assumptions for this first exercise are moderate growing investment in solar projects and, as a possible substitution, a moderate increase in liquefied natural gas (LNG) prices. In addition, this scenario does not incorporate the development of big hydroelectric projects in the south of the country.

The following three scenarios are modifications of part of the assumptions of the first scenario, trying to identify the effect of a particular policy.

The second scenario considers a decrease in the cost of investment in solar projects, with the rest of the scenario's assumptions remaining as the baseline. This assumption is consistent with evidence in the last report of renewable costs by the International Renewable Energy Agency (IRENA). This report states that the falling costs of solar and onshore wind are significant. Solar PV technology experienced learning rates of 18% to 22%, with solar panel prices falling by around 80% since 2010. At the same time, onshore wind shows learning rates of 15%, with costs falling by around 38% since 2009. Taking into account that Chile is a small and open economy, we assume that these prices will be reflected in the local economy as in the global economy.

The third scenario consists of testing the possible expansion of hydroelectric projects in the south of Chile. Because of its geographic location, Chile has a significant hydroelectric potential. In 2015, the most recent survey from the University of Chile (Jiménez-Estevez *et al.*, 2015) established that the total potential is 7.704 MW distributed across 142 generation plants, with about one-third of them in the south of the country. This was calculated by considering operational plants, plants under construction, and provisional plants.

The fourth scenario modifies the last assumption on LNG prices. This scenario is justified because of the dynamic experienced in the industry over the last ten years. This dynamic is characterised by the expansion in US gas production. In the following years, it expects an annual production growth of 1.6%, satisfying the demand of emerging countries like China, which represents 40% of the total demand. The USA is expected to fulfil more than one-third of the extra production in the following five years, making a considerable impact in the natural gas market.

These four scenarios are represented by a baseline that will vary according the level of carbon tax simulated in each scenario. The carbon tax varies from $5 to $50; for simplicity we will report variations of each baseline scenario for $5, $20 and $40 per ton.

Research process and methods

To diagnose the bigger picture of the Chilean energy sector and the resulting multi-level challenges, we conducted open-ended interviews with the main stakeholders in the three main industries: the electric sector, mining, and manufacturing.

We used Mitigation Action Plans and Scenarios (MAPS) scenarios for estimating energy poverty under different future characteristics of the energy sector (MAPS Chile, 2014). These scenarios were built through several rounds of consultation with a mixed group of more than 300 experts from seven sectors: (1) generating and refining transformation; (2) mining and other industries; (3) transport; (4) commercial, residential, and public energy consumption; (5) land use and agriculture; (6) forestry; and (7) waste.

At the same time, the quantitative assessment process was conducted in two phases. The first was about building the baseline, imagining Chile in a business-as-usual scenario and forecasting the economy over 20 years from 2006. The second phase was characterised by the identification of possible measures that could be introduced in the following years in the sectors of Chile's economy. For our proposes, the electricity sector was defined as one of the most dynamic sectors in the adoption of new technologies for electric generation, for example liquefied natural gas, solar, and wind. Experts brought not only technical arguments for consideration in the scenario, but also normative arguments for their justification. This phase was the most significant for considering our estimates at the time for the energy poverty scenarios.

With the possible scenarios identified, we used a minimum cost model representing the electricity generation market in order to obtain the equilibrium price of that electric generation market for each scenario. With that equilibrium price established, we used a macroeconomic model that has a minimal minor representation of the energy sector as part of the inputs of the production function of the economy (together with capital investment and employment) – in other words the total added value. When introducing those equilibrium prices we estimate the output where feedback leads the demand side in the electric model. As outcomes of the interaction between both models, we can see the consumption and investment of a representative household, the deviation of output during the price increase, and the GHG emissions associated with each scenario in the electric model.

Energy poverty as a consequence of the carbon dioxide tax

The current Chilean carbon dioxide tax will give raise to fuel poverty under all scenarios conceived by experts (Figure 6.1). Around 15.7% of households in Chile are in energy poverty in the baseline scenario without the change in prices. Under the four mitigation scenarios this number could increase to 16.19% of the population, controlling for the change in the mix of electricity generation and the different levels of carbon tax simulated.

At the same time an estimated monetary compensation amount for those households could provide each vulnerable household with the amount of money that satisfies their utility function and maintains their level of wellbeing at the level they experienced before the introduction of each mitigation scenario.

Public spending of between US$9.3 and US$15.5 million is required to alleviate the impact of the mitigation scenarios.

Thus energy poverty can be considered a consequential risk resulting from Chile's low-carbon pathway as manifested in this case as carbon pricing in the form of a tax to increase renewable energy in the electricity sector. This risk is observed in two dimensions: (1) the quantitative dimension, resulting from the exercise presented above, using simulations and calculating the monetary compensation required to restore at least at the same level of utility vulnerable households after the carbon tax; and (2) the qualitative dimension where negative impacts in the wellbeing of households further than monetary ones are observed and could affect the comfort of people that in extreme could provoke death.

For the quantitative dimension, it is important to consider that the rich discussion with stakeholders allowed us to work with realistic scenarios of technology penetration and to draw a possible evolution of the carbon tax rate in the midterm. One of the main points to consider was how future increases to the actual rate of carbon tax were to cope with the negative externality and the rate of emissions that Chile will face in the following years.

Certainly, and like the definition of risk in the literature has done, the energy-poverty negative consequential risk depends on the context of the assumptions made in each scenario. A good example of this dependency is the exogenous performance of prices in liquified natural gas (LNG); given that

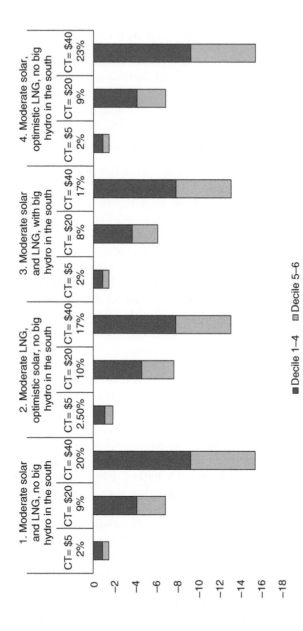

Figure 6.1 Chile's fiscal cost by percentage of variation of electricity prices, decile covered, carbon tax implemented, and scenario of mitigation.

Source: authors' calculations.

Note

Carbon tax (CT) and the electricity price variation.

Chile is a net importer of natural gas prices determined in the global market, this will definitely improve or deteriorate the consequences in energy poverty.

Nevertheless, there are some policy options in the context of the estimations that depend on the willingness of policymakers to incentivise the penetration of other technologies, like big hydropower in the south, resulting in a minor impact in the monetary loss for households.

In the qualitative dimension, according to the suggested energy poverty indicators, a definition of poverty in terms of energy satisfaction, or the ability to maintain a comfortable indoor temperature, is associated with the last perceptions of each individual in the previous season, according of the panel of stakeholders at Red de la Pobreza. For example, a measure suggested by the stakeholders are questions such as, 'How cold for you was the last winter?" or "How hot was the last summer?" The huge variance in possible answers will relate to the observed average temperature and is expected to be both high and not easily comparable.

However, stakeholders consider that these kinds of definitions must be collected in some way in order to introduce them into the design of the policy, mainly because of its negative impact on health and in extreme as a cause of death.

Additionally, different climatic regions in Chile will require different adjustments in indoor temperature. In the southern region, biomass (direct combustion) is traditionally used as heating in the winter. The biomass is readily available and low cost compared with the cleaner heating sources including solar power and hydropower. However, hydropower is not yet fully developed in the south and there are divided views on the development of hydro due to climate change and irregular rainfalls. The potential switch to cleaner energy would increase the cost of heating in these homes but may impact different families in different regions in Chile.

Beyond the two dimensions explained before, one means to reduce the consequential risk would be to recycle revenues from the carbon tax to support the families that are at risk of energy poverty. However, this must be carried out carefully as there are different energy poverty indicators that consider different parameters. For instance, some energy poverty indicators stipulate households are in energy poverty if they spend more than 10% if their income on fuel, but these indicators do not consider household make-up and their energy needs, nor energy infrastructure. Other indicators such as the minimum income standard (MIS) based indicator, will categorise a household as at risk of energy poverty if, after covering its housing requirements, its residual income cannot meet the household's energy costs and other living costs as well as maintain a decent living standard. Another indicator, the low-income–high-cost (LIHC) indicator, considers households to be in energy poverty if they have energy costs that are above the median level.

Different indicators will tell a different story related to the negative outcome of a certain electricity mix and the impact on energy poverty. The first two indicators could show a high risk of energy poverty for the poorest in the country, while LIHC could show that the middle class has a higher risk of fuel poverty. The variations are due to the different costs associated with different energy fuels. For instance, households spend less on low-cost biomass in southern regions versus higher costs of solar power in other regions (see Gonzales, 2018). Currently in Chile, there are no considerations to assess fuel poverty within the different thermal zones and different household income levels. More studies are needed in order to explore the impact of renewable electricity on energy poverty as the climate continues to change and more households will need to adjust to the changes. The burden of the costs will need to be carefully considered and the risks of energy poverty mitigated before rolling out policies.

In summary, successful implementation of a low-carbon pathway will place pressure on household budgets, as energy costs will rise. Any of the possible scenarios would change both actual and desired spending decisions on energy and other essential items in individual households – what we call their 'preferences'. Many household preferences cannot be directly observed, but we consider that an economic compensation, set by observed expenditure on energy products, is the best proxy for the hidden preferences of these households. This measure will not fit all the preferences of every household but it will at least cover a significant proportion of them.

Other uncertainties and risks

The policy complexity at the intersection of economics, energy, and environment may give rise to uncertainties, and under these some identifiable risks related to the promotion of renewable energy in Chile which go beyond the consequential risk posed by energy poverty. They have not been quantified but are the result of stakeholder engagement processes including focus groups.

We understand the uncertainty as a lack of information about possible results or stages after policy implementation. In this case the uncertainty, identified by the stakeholders' discussion around scenarios of mitigation for a low-emissions economy, is nested in the behaviour of citizens in cultural and social demands towards the promotion of renewable energy.

This unpredictable behaviour of the citizens regarding the promotion of renewable energy is contrasted by actual positive positions to renewal technologies in regions far away from solar or wind farms, and contrasted by the opposition observed in some regions of the country where they host these farms. A good example of this contradictory behaviour is the opposition to wind farms in the south of Chile, specifically in Chiloe. This kind of manifestation opens the door for doubt about the performance of the renewal sector in the future. While there is an increasing willingness to adopt new and clean technologies, other kinds of considerations, like the beauty and quality of landscapes, also play an important role in the subjective valuations of society.

At the same time, a cultural change also brings uncertainty around the feasibility of the promotion of the renewable energy as an instrument of alleviating negative externalities in climate change policy. A good example of this is the treatment of biomass or wood in the south of Chile. Not only monetary con-cerns are considered for the appositions of the eradication of this kind of source of energy, but also all the traditional and cultural conceptions of wood in the living standard of the region play a major role in the introduction of new clean technologies.

The uncertainties identified above bring us to a subset of implementation risk, that is the expansion of electricity access across Chile as a heating and cooking option. Currently in Chile the coverage of electricity access is 99.6%. The remaining ~0.4% is spread around the country. From this data we can appreciate that the largest portion of the 0.4% without electricity connection is concentrated in one of the poorest regions of the country, the Araucanía region. Furthermore, trying to characterise this region, we can learn from Encuenta Nacional de Energia (Ministerio de Energía de Chile, 2016) that the Araucanía region uses 70% biomass-wood for heating and less than 5% electricity for this propose.

One of the implementations risks in this context is related to the monetary expected change for citizens. There is no official market for biomass-wood in Chile and prices are lower than the estimated prices of electricity if this option is considered as an option for heating. Also, it is important to consider the cost of these alternatives in the building and operational phases, and to consider that one of the natural constraints of the energy matrix in Chile is the dependency on fuel imports (IEA, 2018).

Another risk is observed at the time of subjective considerations related to renewable energy and energy poverty. According to an expert's panel named Red de Pobreza Energetica, more significant than the monetary dimension of the consequential risk of energy poverty, they observed that energy poverty could have three negative impacts in the lives of citizens through these dimen-sions (Urquiza *et al.*, 2017).

The first impact observed is the socio-cultural dimension. According to the discussion of this group reported in Amigo *et al.* (2018), the importance of abso-lute and relative energy necessities is differentiated by the diversity of regions of Chile demanding this feature in the policy design.

The second impact is determined by the first one and is related to the quality of energy access in regions. Factors like equipment, housing conditions, and the intermittency of the service are part of this impact. In this regard, preliminary observation of expert judgement of the group estimated that the impact would be observed in the monetary budget of the household.

The third impact is with respect to accessibility and equity. The most signi-ficant challenge in this regard is the conditions of towns at the borders of the country where energy sources are not available and where the provision distri-bution is costly. This third dimension is closely related with the consequential risk analysed in the previous sections by quantitative methods.

Conclusions

Chile is facing new and complex demands as a result of its economic progress. One of those demands is moving to a low-carbon economy while keeping economic growth on the same path. For this purpose, the government actively promotes renewable energy to reduce GHG emissions, relying on the significant potential of renewable energy in Chile. In order to support the diffusion of the respective technologies with market instruments, the government of Chile recently introduced a carbon tax.

This tax, however, will create an unintended collateral problem – a consequential risk – affecting disproportionally the most vulnerable portions of the population. In a scenario where levels of penetration of solar technologies for the electric generation interacts in four different mixes of energy, 15.7% of Chilean households are in energy poverty. Without economic compensation and with a variation of carbon tax from $5–40/ton of CO_2, it would lead to a 16.2% increase in the proportion of households in energy poverty. If the government decided to compensate the most vulnerable in the population, the cost would be between US$9.3 and US$15.5 million for a programme that left households at at least the same level of wellbeing and comfort as before the tax.

Apart from the risk of increasing energy poverty, we identify an uncertainty of the behaviour of citizens towards renewable energy promotion. This emerges after a rich exchange of ideas with stakeholders at the point of building the scenarios. Moreover, we identify implementation risks that at the moment are not quantifiable because of the lack of detailed information at household level, these being the monetary cost of substituting biomass-wood by electricity if this is considered an option for heating at household level, and the multidimensional risk in the mix of quantitative and qualitative perceptions of the stakeholders can be seen (see Figure 6.2).[2]

Notes

1 Researcher in Tenaris until February 2018.
2 We would like to thank Hernan de Solminihac and Leonardo Hernandez (director and co-director of Clapes UC at Pontificia Universidad Católica de Chile); Felipe Larrain (Minister of Finance of Chile and former director of Clapes UC), Luis Abdón Cifuentes, Carmen Gloria Contreras, Jenny Magger, Marek Antosiewicz, and Sebastian Vicuña, presenters in the second seminar of Clapes UC and TRANSrisk 'Medición y gestion del cambio climático'; the entire team of Red de Pobreza Energética; Luis Cuervo-Spottorno (Primer Consejero of the European Union delegation in Chile), with Alejandro Zurita (Chief of Science, Technology and Innovation at Servicio Europeo de Acción Exterior of the European Union) and Leticia Celador (trade and economic adviser at the European Commission delegation in Chile).

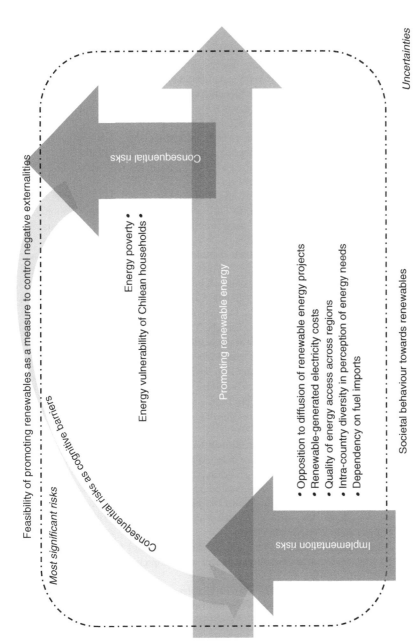

Figure 6.2 Main uncertainties, implementation, and consequential risks associated with a pathway driven by renewable energy expansion in Chile.

References

Amigo, C., Araya, P., Billi, M., Calvo, R., Oyarzún, T., Urquiza, A. (2018). *Políticas públicas y pobreza energética en Chile: ¿una relación fragmentada? Documento de Trabajo Red de Pobreza Energetica.*

Benavente, C., Gonzales, L., Díaz, M. (2014). The Impact of a Carbon Tax on the Chilean Electricity Generation Sector. *Energies* 8, 2674–2700.

Generadores de Chile (2018). *Gobierno y Generadoras anuncian el fin de nuevos desarrollos de plantas a carbon.* Available at: http://generadoras.cl/media/page-files/391/180129%20Comunicado%20no%20mas%20nuevas%20plantas%20a%20carb%C3%B3n%20-%20ME%20MMA%20Generadoras%20de%20Chile.pdf.

Gobierno de Chile (2015). *Ministerio de Medio Ambiente de Chile. Contribución Nacional Tentativa de Chile (INDC) para el Acuerdo Climático París 2015.* Available at: www4.unfccc.int/Submissions/INDC/Published%20Documents/Chile/1/Chile%20INDC%20FINAL.pdf [accessed 16 November 2016].

Gonzales, C.L. (2018). Energy Economy and Welfare Effects. Working paper Pontificia Universidad Católica de Chile, Centro LatinoAmericano de Políticas Económicas y Sociales ClapesUC. Available at: www.dropbox.com/s/r2qje2qorr521jp/Energy%20Poverty%202018%20Final.pdf?dl=0.

IEA (2018). *Energy Policies Beyond IEA countries, Chile 2018.* International Energy Agency. Available at: www.iea.org/newsroom/news/2017/march/iea-finds-co2-emissions-flat-for-third-straight-year-even-as-global-economy-grew.html.

IPCC (Intergovernmental Panel on Climate Change) (2014). Summary for policy-makers, in: Field, C.B., Barros, V.R., Dokken, D.J., Mach, K.J., Mastrandrea, M.D., Bilir, T.E., Chatterjee, M., Ebi, K.L., Estrada, Y.O., Genova, R.C., Girma, B., Kissel, E.S., Levy, A.N., MacCracken, S., Mastrandrea, P.R., White, L.L. (Eds.), *Climate Change 2014: Impacts, Adaptation, and Vulnerability. Part A: Global and Sectoral Aspects. Contribution of Working Group II to the Fifth Assessment Report of the Intergovernmental Panel on Climate Change].* Cambridge University Press, Cambridge, United Kingdom and New York, NY, USA, pp. 1–32.

IRENA (International Renewable Energy Agency) (n.d.) *Dashboard.* Available at: http://resourceirena.irena.org/gateway/dashboard/

Jiménez-Estevez, G., Palma-Behnke, R., Roman Latorre, R., Moran, L. (2015). *Heat and Dust: The Solar Energy Challenge in Chile.* Available at: http://repositorio.uchile.cl/bitstream/handle/2250/133141/Heat-and-dust.pdf;sequence=1.

Ley No 20257 (2008). *Ley que introduce modificaciones a la ley general de servicios eléctricos respecto de la generación de energía eléctrica con fuentes de energías renovables no convencionales.* Available at: http://centralenergia.cl/uploads/2009/12/Ley_ERNC_LEY-20257.pdf [accessed 28 November 2016].

MAPS Chile (Mitigation Action Plans and Scenarios for Chile) (2014). *Resultados Fase 2, MAPS Chile Gobierno de Chile.* Available at: www.mapschile.cl/files/Resultados_de_Fase_2_mapschile_2910.pdf.

Ministerio de Desarrollo Social (2017). *Situación de Pobreza.* Available at: http://observatorio.ministeriodesarrollosocial.gob.cl/casen-multidimensional/casen/docs/Resultados_pobreza_Casen_2017.pdf [accessed 27 August 2018].

Ministerio de Energía de Chile (2016). *Encuesta Nacional de Energía.* Santiago de Chile.

Urquiza, A., Amigo, C., Billi, M., Leal, T. (2017). *Pobreza energética en Chile: ¿Un problema invisible? Análisis de las fuentes secundarias disponibles de alcance nacional.* Publicaciones REDPE Universidad de Chile.

7 The Netherlands

Expanding solar PV – risks and uncertainties associated with small- and large-scale options

Krisztina de Bruyn-Szendrei, Wytze van der Gaast, and Eise Spijker[1]

Introduction

According to recent projections (ECN, 2017), the Netherlands will not be able to comply with its commitment under the 2008 Energy and Climate Package to realise a 14% share for renewable energy consumption by 2020. Currently, only about 6% of energy is generated from renewable sources, which is expected to increase to, at most, 13% by 2020 (European Commission, 2017). In comparison, in several other EU member states – such as Bulgaria, the Czech Republic, Denmark, Estonia, Finland, Italy, Croatia, Latvia, Romania, and Sweden – the 2020 targets have already been achieved. Other member states have not yet done so but are on track. Next to the Netherlands, France and Luxemburg are other member states that have not yet complied, nor are on track for realising 2020 renewable energy targets (European Commission, 2017).

In response to the slow progress, a policy package was prepared by the Netherlands Ministry of Economic Affairs in 2016 with additional measures for reaching the 14% target by 2020 (Kamp, 2016). While technically the options in the package seem feasible, there are a range of barriers affecting their successful implementation, such as costs, spatial planning, and public acceptance issues.

To date, biomass (63% in 2016) and wind energy (24% in 2016) have been responsible for the bulk of renewable energy production in the Netherlands (CBS, 2017). Solar energy constitutes 5.4% of renewable energy in the country, which is not a surprise given the relatively low solar radiation in this region of Europe. Nevertheless, due to the ongoing debate on the sustainability of (imported) biomass and the increasing difficulties related to expanding onshore wind power in the Netherlands (RVO, 2017), expanding solar energy on rooftops as well as on land and water has increasingly become an indispensable option for complying with renewable energy commitments for 2020 and beyond.

Solar energy can provide different energy services, such as for electricity, heating, cooling, and transportation. Each of these services face challenges and therefore risks, such as space needed for solar parks, extra costs to be borne when purchasing and operating the panels, and social aspects related to energy switch and acceptance of the technology. This chapter focuses on the possible consequences of expanding electricity production through solar energy. This has several common aspects and potential risks with other renewable energy sources but faces additional risks related to grid balancing.

Despite their current small share of solar-based power production in the Netherlands, ground-mounted parks could potentially become at least as important as rooftop PV (DNV GL, 2016). Therefore, in this chapter, both technology options for solar PV are considered for a pathway towards expansion of electricity production in the Netherlands through rooftop solar panels and large-scale solar parks. Possible consequences (risks) of this pathway for society and the economy are elaborated, as well as potential barriers that could prevent successful scaling up of the technology options in the country.

The Dutch pathway towards solar PV

In October 2017, the newly elected Netherlands coalition government announced a national climate and energy accord with the goal of reducing emissions of greenhouse gases (GHGs) by 49% by 2030 (below 1990 levels) (VVD, CDA, D66, ChristenUnie, 2017). This goal corresponds with an emission reduction of 56 Mton CO_2, on top of reductions through existing and planned policies for mitigation in the Netherlands. Eighteen Mton of that reduction will be obtained through the accelerated phase out of coal plants, which means that alternative electricity sources need to be utilised, including solar PV.

With a view to the recent policy context in the Netherlands, this chapter analyses potential consequences and implementation barriers related to a desired future with large-scale application of solar PV for reaching Dutch and European energy and climate goals, as well as a trajectory with actions to realise that (a pathway). As part of the pathway, we consider two options: (1) the upscaling of rooftop solar panel use in the built environment with a focus on households, small businesses, schools, etc.; and (2) large-scale applications of solar PV, such as through ground-mounted solar parks of high output.

The starting point for the pathway is that, presently, the installed solar-power capacity of the Netherlands is $2.7\,GW_p$ (CBS, 2018), 94% of which is based on rooftop solar panels on dwellings and buildings; the remainder is based on ground-mounted solar parks ($176\,MW_p$) (Zon op Kaart, 2018). According to the National Solar Power Action Plan for 2016 (DNV GL, 2016), 4 GWp of installed capacity could be achieved by 2020 (about 12 PJ of generated solar power, assuming 850 load hours). Technically, however, the potential is much bigger. Nowadays, only about 450,000 dwellings in the Netherlands are equipped with solar PV rooftop panels, while potentially around four million roofs are suitable for that (which is around half of the total number of dwellings in the Netherlands). Should this technical potential be utilised, in addition to the roofs of industrial and utility buildings, solar PV capacity could grow to 70 GWp in 2075 (DNV GL, 2016). Fully utilising the technical potential of large-scale solar parks could double this capacity. Large-scale projects have thus far been scarce in the Netherlands, which has been largely due to the fact that revenues have long been insufficient to cover the costs. Moreover, in a densely populated country such as the Netherlands, space is limited so that opportunity costs are relatively high (see also elsewhere in this chapter when discussing specific risks). For these reasons, according to the chair of Holland Solar (a solar PV sector organisation) in an interview, stakeholders such as municipalities, railway infrastructure management, and the Ministry of Infrastructure and Water Management have long taken a re-active position, while they have only recently become pro-active (*Cobouw*, 2018).

Since 2017, the option of large-scale solar parks has become eligible for exploitation support under the Dutch sustainable energy subsidy scheme (SDE+).[2] Projects granted this subsidy become financially viable, which takes away an important financial barrier to large-scale solar expansion. In 2017, over half of the SDE+ budget of €6 billion was allocated to solar parks (Van den Eerenbeemt, 2018). With the adoption of large-scale ground-mounted projects, the expansion of solar PV could go faster than rooftop solar PV due to economies of scale.

Next to technical and economic potential (supported by incentive schemes), the scale and pace of solar PV expansion also depends on a range of other factors. These include: the policy mix for creating an enabling environment for solar PV; opportunity costs when land that is used for solar parks can no longer be used for other (e.g. agricultural) purposes; inefficiencies in the market value chain; and whether energy, legal, financial, and technical services are sufficiently capable of supporting scaling up solar PV.

The scope for Dutch energy and climate policies is set by EU directives such as the Renewable Energy Directive, the Energy Performance of Buildings Directive, and the Electricity Market Directive. Examples of policy instruments derived from these, with direct relevance for solar PV, are: net metering, enabling households and small businesses[3] to feed surplus solar electricity into the grid; the SDE+ subsidy scheme; energy investment tax exemptions, enabling an investor in clean energy technology to deduct investment costs from income before taxes; and the postal code regulation, enabling consumers to invest in nearby solar parks and receive tax rebates.

In particular, the net-metering instrument has turned out to be a strong stimulus for rooftop solar PV investments by households and small businesses. Households that produce solar-based electricity are exempted from energy charges and are paid by the grid operator when they deliver surplus electricity to the grid. With these financial benefits, return on investment for solar PV easily surpassed interest rates on bank accounts. As explained above, the SDE+ opening for large-scale solar PV has generated a strong growth in planned solar parks. Since the eligibility of solar parks under SDE+, the number of completed parks has grown from $2\,MW_p$ in 2014, to $15\,MW_p$ in 2015, $43\,MW_p$ in 2016 and $114\,MW_p$ in 2017. Many more projects are in preparation: as of May 2018, 94.5% of all planned solar parks (including those that have been approved for

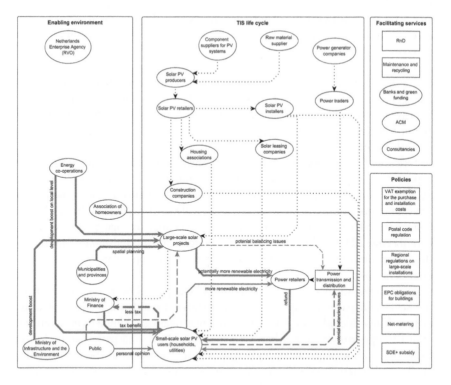

Figure 7.1 The market system map for solar PV expansion in the Netherlands.

Note

The block arrows in the map show connections between institutions and stakeholders that have a positive effect on scaling up solar PV. The broken line arrows reflect potential market barriers, such as public resistance to large-scale solar projects or grid balancing issues due to households' guaranteed delivery of domestically produced solar-based electricity to the grid.

the SDE+) are still in the process of preparation, such as arranging permission and other planning (Zon op Kaart, 2018).

Further promotion of the pathway takes place via (governmental) institutions that develop visions for energy and climate, have responsibility for (spatial) planning, and formulate policy packages around these. Municipalities, for instance, usually issue the permits for parks, for which they need to decide whether a solar park is in accordance with local spatial plans. The ministries are responsible for the national energy and climate policy development, including incentive schemes such as subsidies and tax exemptions, and are thereby supported by government agencies such as the Netherlands Enterprise Agency (RVO). The market system for expansion of solar PV in the Netherlands is shown in Figure 7.1.

Risk elicitation

While the pathway promotes expansion of solar PV, there are multiple factors that can positively or negatively influence its implementation. These are likely to differ between small and larger-scale applications. When taking a decision on pursuing a pathway, it is important to understand both the potential consequences of the pathway and the implementation barriers in terms of how easily, and against what costs, these can be cleared. Therefore, key questions answered in this chapter are: what are the possible consequences of, and implementation barriers to, rooftop or ground-mounted solar PV systems? How are these reflected by risk profiles of the pathway? And how can these risks be mitigated?[4]

Research process and methods

To identify the implementation and consequential risks of solar PV acceleration in the Netherlands, a literature review was conducted which has been supported by a series of semi-structured qualitative interviews during 2016 and 2017 with 19 relevant public and private stakeholders in different relevant sectors and professions. The initial list of the expert stakeholders was put together based on the prior knowledge of team members (purposive sampling). The list was later expanded by adding more expert stakeholders based on inputs from early interview subjects (snowball sampling). Nineteen experts have been interviewed, who were selected from seven stakeholder groups: public body ($n=5$), businesses ($n=3$), academia ($n=3$), energy think tanks ($n=3$), grid/network operators ($n=2$), an energy co-operative ($n=1$), and an environmental non-governmental organisation ($n=1$). The interviewees' expertise spans from technical design and installing of solar PVs to economic, behavioural, and spatial aspects of low-emission transitions. Stakeholders were asked about their views on the transition pathways and policy instruments, implementation barriers, consequential risks, and recommended changes to address these risks.

The interviewed stakeholders have different interests regarding solar PV implementation. For instance, while a policymaker at the national level may strive for achieving renewable energy and climate goals, a local policymaker needs to accommodate relevant measures in local or regional spatial planning. Network operators have a key interest in maintaining grid stability with accompanying resources. Research institutes have a broad interest in understanding problems related to pursuing energy and climate goals and formulating solutions that serve all parties involved in the energy transition. Environmental NGOs interviewed expressed mainly concerns about how low-emission energy and climate options could be hampered by policy, economic, and social obstacles. On the commercial side, interviewed business stakeholders explained their experiences with setting up solar parks and the opportunities and limitations to that. Their experience resulted in often practical issues, such whether and how to clean panels in a solar park, responsibility for connecting to the grid, what professionals are usually asked in order to construct a park, etc. We also included in our stakeholder assessment representatives of an energy co-operative who, from a more ideological perspective, strive for locally generated clean energy.

Large-scale deployment of rooftop solar PV and ground-mounted solar parks is likely to have different types of impact, leading to different types of risk. Similarly, as both technology options in the pathway have different implementation characteristics (rooftops versus ground-mounted investments), their implementation risks are likely to be different too. In this section, key consequential and implementation risks are highlighted (see Figure 7.2 for a summary), as well as ways to mitigate these. As explained in Chapter 2 and illustrated in Figure 7.2, both risk categories are intertwined with each other. For instance, should stakeholders be aware of a potential negative impact of a pathway, such as employment losses, and consider this a consequential risk, then this could also result in a reduced public acceptance, which in its turn would lead to an implementation risk (in Figure 7.2 this is referred to as 'consequential risk as cognitive barrier'). Where applicable, such cognitive barriers are also addressed.

While applying the same type of technology, the small and large-scale pathway options are very different in terms of project size (e.g. small-scale rooftop PV panels versus large-scale ground-mounted solar parks), grid stability measures, ownership, costs, etc. Implementation risks are therefore separately discussed for both options.

Implementation risks related to expanding rooftop solar PV in the Netherlands

Technical aspects

For several reasons, not all residential dwellings and industrial buildings are suitable for placing solar panels on their rooftops. Ideally, a roof is sufficiently strong to carry the weight of the solar panels, has a favourable orientation towards the sun, while the waterproof layer on the roof tiles is suitable for placing solar panels. Lack of these conditions can lead to implementation problems which entail the risk that the desired scale of solar PV for meeting Dutch renewable energy targets is not met. Moreover, even if these conditions are met, owners of the dwelling or building may still find the investment risky or unattractive because, for example, they find panels unattractive, cannot cover the costs, or are concerned about the payback time. These risks are further explained below based on inputs from stakeholders.

Industrial buildings generally have flat rooftops and are generally constructed to support one metre of snow. Such constructions may not be strong enough to carry a layer of solar panels and additional investments to strengthen the roof may be required. Traditionally, most rooftop solar panels are installed on rooftops with a southward orientation for better utilisation of solar radiation, as these have the most attractive business cases. However, the business cases for east- or westward-oriented buildings have recently improved due to technical improvements with enhanced efficiency and the net-metering policy. Yet the latter category seems to be more sensitive to a change in this policy: one of the interviewed researchers indicated that, should net metering be abolished,

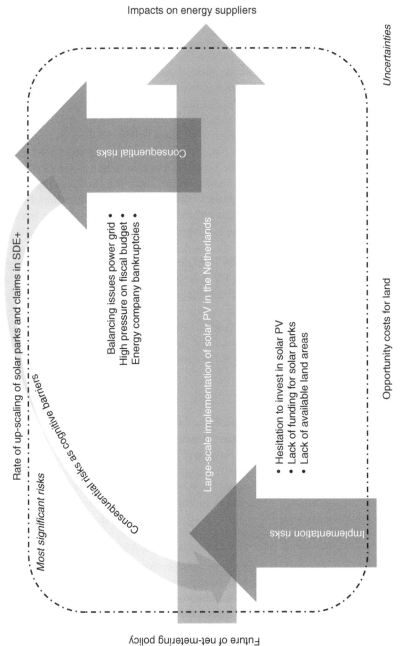

Figure 7.2 Main uncertainties, implementation, and consequential risks associated with a solar PV-powered transition pathway in the Netherlands.

the payback time of southward-oriented panels would increase from five to seven years, while the payback time of east- or westward-oriented panels increases from nine to 15 years.

Social aspects

From the perspective of social acceptance of rooftop solar panels, the survey conducted among Dutch stakeholders has shown that public perception of rooftop solar PV is generally positive. An important enabling factor to that has been the policy instrument of net metering,[5] which provides a good business case for owner-occupiers who have both a sufficiently large rooftop area and the financial ability to invest in solar PV. A considerable proportion of Dutch households, however, do not reap the benefits from this policy as they consider the payback time to be too long or simply do not have the funds necessary for the initial investment. Partly this relates to the argument that the rationality of people is hampered by not having full information as well as cognitive limitations, while time for decision making is limited (the concept of bounded rationality as explained in Chapter 2).

For example, while a comparison between interest revenues from savings accounts and revenues from solar PV investments may currently lead to a favourable picture for the latter, households may still find it more comfortable (a lower perceived risk) to keep their money in a savings account. This is partly due to the underlying uncertainties, such as 'unknown' developments of known risks of household investments in solar PV. In the case of investment in solar PV panels, this could lead to postponed or cancelled investments.

Policy-related uncertainty

Finally, uncertainty for scaling up rooftop solar panels, which could result in an implementation risk, lies in the current absence of a policy to make newly built residential dwellings more suitable for solar PV panels. According to the consulted stakeholders, it would be essential to provide incentives or regulation so that architects and building companies, as well as future home owners, can better integrate solar panels on rooftops and/or situate houses in such a way that it is most effective for solar PV. Currently, the energy performance code (EPC) for buildings offers insufficient incentives for this, so newly built houses may still lack suitable conditions for solar PV, and instead stimulates doing the minimum of installing only two panels. (These are also referred to as 'shame or excuse panels', according to the interviewed solar PV experts.) For existing houses this is even more complicated because a building certificate (an energy label from A = 'highest' to G = 'lowest') is only required if a dwelling is sold.

With stricter EPCs, installing solar PVs would be a relatively cheap way to meet the new standards. Adopting solar PVs on rooftops could therefore become mandatory in building standards and guide architects when designing new houses.

Implementation risks related to ground-mounted parks for solar PV

Public resistance

Ground-mounted solar parks will, similarly to onshore wind turbines, have an impact on landscapes as they require a significant piece of land and are often clearly visible from long distances. In the Netherlands, spatial plans developed by provinces and municipalities determine for what purposes land can be used. This could give rise to social conflicts if a solar park is eligible for a certain area according to the spatial planning but is not preferred for this area by the local population.

Information evenings held for ground-mounted solar projects in the development phase showed that the main concern of the local population is where the park is going to be located and how this relates to the location of people's houses. There is a preference for placing solar parks outside cities where they are less clearly visible. This contradicts with, for instance, the regional regulations in the Province of Fryslân where, according to spatial planning, ground-mounted solar parks can only be installed near cities/villages and not in open land (Gemeente Súdwest-Fryslân, 2015). The spatial departments of provinces often oppose having solar parks in the countryside because, during the past 30 years, their primary task has been to protect the rural landscape. These contradictions between traditional spatial planning and people's preferences regarding location of ground-mounted solar parks could considerably delay the deployment of solar parks due to lack of acceptance (or even resistance) among the local population.

Therefore, an important lesson from the stakeholder consultation for this study is that it is important to engage the local population as soon as possible in decision making on a ground-mounted solar park. This is important as a solar park can, as explained above, affect the population in multiple ways, such as visual pollution due to the significant change in the landscape, opportunity costs as the land can no longer be used for other purposes for a long period of time, and associated issues such as reduced job opportunities. Recently, in the Netherlands, there have been cases where ground-mounted solar parks were cancelled due to local resistance and in these cases the population, as an important stakeholder, had only late in the process been consulted. Consulted stakeholders for this study have therefore identified social resistance as a key obstacle to scaling up solar PV through ground-mounted parks.

According to interviewed commercial parties, a way to lower public resistance against solar parks can be found in improvements in the design and (spatial) planning structure. Currently, most project cases focus on cost optimisation as that is the key criterion for obtaining subsidies through the SDE+, so that project developers have an incentive to choose relatively low-cost materials even if this has a less attractive design. Moreover, as explained previously, solar parks in the Netherlands fit best in areas close to population centres as using more remote, rural areas would require modification of spatial planning.

Therefore, consulted stakeholders recommend that the design of the panels be considered as a factor when taking a decision on providing subsidies, and that spatial plans are modified so that solar parks are also eligible for rural areas.

An example of a concept where social acceptance is more naturally considered in the project planning is that of energy co-operatives, which are generally formed by locals who are familiar with regional issues and preferences. Trust in energy co-operatives is therefore higher than in government bodies because they offer the opportunity for locals to be part of the decision making on preference for energy source, to have a voice, to be critical, and to raise questions regarding their spatial environment.

Further to risks related to implementation of both solar PV options and how these may block or slow down expansion of solar PV in the Netherlands, this section focuses on potentially negative impacts of the options (consequential risks) for economic sectors and areas of society. Contrary to the former section, where implementation risks are discussed separately for scaling up rooftop solar PV and ground-mounted solar parks, the potential consequential risks that follow are discussed mostly simultaneously for both options as the risks are largely similar. Where differences in consequential risks exist, these are explicitly mentioned.

Consequential risks of expanding rooftop solar PV and solar parks in the Netherlands

Impact on the electricity grid

Presently, the Dutch electricity grid is designed to handle peak demand but, with an increased share of distributed renewable energy such as solar PV, the current peak handling capacity will become insufficient or at least come under pressure. Distribution grids are sized around average demands of 1 kW per dwelling and become insufficient if (instead) dwellings are feeding 4–8 kW (peak) into the grid (DSO, 2017). Technically, modifying the grid is not a complex task but there will be costs involved, which are covered by the distribution grid operator. As these costs do not directly accrue to solar PV panel owners, they are usually not considered by households when deciding on whether or not to invest in rooftop panels.

Grid operators are not involved in small-scale rooftop solar PV projects and therefore do not have influence on where and how many panels are installed. If all suitable homes (around four million, as discussed earlier) were equipped with solar panels, about 16 GW of solar power could be generated per year. Operators interviewed in our stakeholder consultation indicated that the grid could cope with this amount of solar power with only small adjustments but, in order to avoid grid-balancing issues, large neighbourhood batteries may be needed to store generated electricity that cannot be used momentarily due to insufficient demand.

From the households' point of view, net metering is an attractive instrument, but from a grid-balancing perspective it is more complex and not the preferred way to stimulate solar installations. The price of solar energy varies throughout the day, while in the system of net metering the price is fixed. There is no incentive to use electricity at times when it would be best for the electricity grid. Ideally, solar electricity is used when available (during sunny hours). During these hours, electricity prices would go down because of increased solar electricity supply. Dynamic electricity pricing could facilitate a smarter way of using electricity so that peaks in supply are met with increased demand and balancing issues are mitigated or avoided. Next to technical investments, this will also require modification of the legal framework that governs the electricity market.

In contrast to individual rooftop solar PV projects, grid operators are generally involved in the development of large-scale solar projects as they need to arrange the necessary infrastructure (connections, cables, etc.). Network operators are obliged to connect the park to the grid when they are called upon to do so. Grid operators interviewed expressed concern that if the recent acceleration in project applications (such as under the SDE+) continues with a high rate of project approval, then in the short term large grid reinforcement, as well as smart grid development, will be necessary. A very recent example concerning the limited grid capacity came from the south-east of Groningen, where the DSO and TSO warned that in the short term they will not be able to connect new solar parks due to the risk of 'electricity congestion' in the local grid (Van de Veen, 2018).

One way to reduce the costs of grid reinforcement, as suggested during the stakeholder consultation for this study, is by placing solar parks in 'grid-friendly zones'. Two-thirds of the grid connection costs related to a new solar park could be avoided by placing solar panels in areas where combinations with existing connections can be established, such as large-scale rooftop projects, or next to wind parks, industrial areas, and railways (because this infrastructure can also be used). According to such a 'smart design' strategy, projects make use of already-existing infrastructure. Innovations like this are, however, hampered as grid operators are generally not considered market entities but are obliged to follow regulations without influence on spatial planning decisions.

An efficient solution, suggested by consulted stakeholders, could be to include a section in the SDE+ subsidy scheme saying that projects that are planned for a grid-friendly zone (following smart design) will receive a preferential treatment and receive more support per kWh. Currently, the SDE+ is designed in such a way that cheaper projects are granted first and generally with a lower tariff per kWh. This disregards the cost for the grid operator.

Impact on traditional electricity producing sectors

Expanding the use of renewable energy at the expense of traditional, fossil-fuel-based energy plants may have a negative impact on the business cases of fossil-fuelled power plants. Coal and gas power plants will most likely be used as flexible power production capacity to accommodate the peaks in daily energy demand. 'Traditional' energy companies can furthermore support the low-emission energy transition by shifting to more renewable-energy production themselves, thereby benefiting from their existing customer base and experience with large-scale projects.

Impact of net metering on the fiscal balance of the Dutch treasury

As explained previously in this chapter, rooftop solar PV panel investments by households are supported via the policy instrument of net metering. Ground-mounted solar parks have become eligible for exploitation support through the SDE+ subsidy scheme. While net metering has an impact on the government's fiscal budget through reduced tax income, subsidy scheme support for solar PV will lead to higher governmental expenditure.

Due to net metering, the Dutch treasury misses potential tax revenues because households and businesses with solar PV panels on their rooftop are exempted from paying energy tax, value-added tax (VAT), and sustainable energy contribution (ODE) on the self-generated electricity they consume. For this study, a simplified model exercise has been carried out to calculate this effect, assuming an accelerated phase out of coal plants by the year 2030.[6] It has been concluded that, should the net-metering mechanism remain unchanged, the treasury might lose out by around €500 million in 2023, rising to almost €1.2 billion by 2030.

However, for a more complete picture, for this study the net financial impact of net metering on the Dutch treasury has been estimated. After all, economic activities related to installing solar panels also lead to tax incomes. Assuming that an average household generates about 3000 kWh of solar electricity per year and is exempted from energy tax, VAT, and ODE, the government's annual tax income amounts to around €1,200 per year (compared to a situation where the household consumed electricity from the grid and paid taxes on this). The calculations show that tax losses for the treasury indeed outweigh the gains in tax income from installation companies (see Figure 7.3).

According to interviewed stakeholders, net metering has been a powerful incentive for households to invest in rooftop solar panels. Additionally, as some indicated, the persistently low interest rate has been a stimulus as this made saving money generally less attractive than investing in solar panels. An

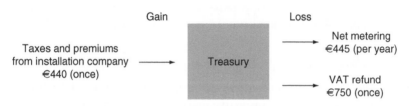

Figure 7.3 Balance of the Dutch treasury for the case of a household with 3000 kWh solar electricity generation.

important condition for the latter has been the trust of households in the continuation of the net-metering policy instrument. Uncertainty about precisely that has recently arisen due to communication by the government about a planned evaluation of the instrument and possible changes based on that. Already this news has created a hesitance to invest in rooftop panels.

With respect to the impact on the government's fiscal budget of subsidising large-scale, ground-mounted solar parks, a similar modelling exercise was undertaken for the Netherlands, assuming different scenarios for phasing out coal-based electricity and scaling up solar PV. In a scenario of an accelerated phase-out of coal, the expenditures related to the SDE+ subsidy in support of solar parks would amount to almost €5 billion per year by 2030; in a business-as-usual scenario, based on recent policies in the Netherlands, these expenditures would amount to around €1 billion per year by 2030.

Impact on employment

Both pathway technology options are expected to generate jobs because a workforce is needed for the design, installation, and maintenance of the panels. Presently, however, there is little to no expertise with the design and construction of large-scale solar PV projects in the Netherlands, so often foreign experts are contracted to oversee the process. In addition, a stakeholder engaged in developing large-scale solar parks in the Netherlands indicated that construction of the parks is often carried out by specialised teams with foreign employees who travel from project to project in Europe. Local employees are mainly contracted for their knowledge of local rules and regulation (electrical, legal, etc.). After the projects are operational, however, only a minimal workforce is needed for operation and maintenance as solar parks usually do not require much cleaning and maintenance.

Discussion of risks and uncertainties related to the solar PV pathway

Part of the case study analysis is to assess how the solar PV low-carbon transition pathway performs in terms of its contribution to meeting the Dutch renewable energy targets and realising other socio-economic benefits. As explained, both technology options are needed for reaching the Dutch renewable energy target of 14% in 2020 (and beyond) and the 2030 climate target set by the government coalition, whereby large-scale solar projects particularly contribute to accelerating renewable energy expansion, mainly because of economies of scale in planning, financing, and construction. Considering overall cost efficiency of public spending and fiscal effects, it is still unclear which pathway is most cost-effective. With ground-mounted solar parks, potential production is relatively high but the subsidy-related expenses are also considerably higher than the reduced tax revenues when supporting rooftop solar PV.

While for the success of the pathway all risk aspects discussed in this chapter will have to be addressed, from the stakeholder consultation it is clear that spatial planning will be among the most important factors for success. For small-scale rooftop applications of solar PV, the challenge is to ensure that newly built dwellings have a better orientation towards the sun. For larger-scale projects, especially ground-mounted solar parks, spatial planning needs to consider alternative uses of areas so that opportunity costs remain low. They should also consider whether an area is acceptable for residents, especially when projects are adjacent to their villages and towns and therefore clearly visible by the local population. From the stakeholder interviews, a clear recommendation is to involve local stakeholders in the decision-making process for ground-mounted parks as early as possible. This helps to make people 'co-owner' of the project and enable them to propose modifications to the project plan. Recent examples of solar-park plans in the northern region of the Netherlands have shown that, if project developers do not take into account public opinion and consider different design options, public opinion might quickly change and prevent the implementation of large-scale projects in the long term. This is supported by findings by Nikas *et al.* (2018) which indicate that 'soft measures' focusing, for instance, on behavioural change and public acceptance can lead to more sustainable low-emission pathways than measures focusing on mainly providing financial incentives.

The findings from the stakeholder consultation moreover suggest that several of the risks discussed in this chapter are augmented by uncertainties about policy development and spatial planning, and a lacking larger picture of the overarching climate and energy perspectives of scaling up solar PV. As a result, optimal investment decisions as perceived by individual stakeholders may not be optimal from macro-level economic, social, energy, and climate perspectives. This calls for a 'grand design' solution or a strategy to 'fix' the grid-balancing, finance, and spatial planning issues. Part of the grand design could be a prioritisation of risks to be tackled first, which would not be driven purely by efficiency of resource allocation but also by cost-effective use of solar PV for realising the Dutch contribution to European energy and climate goals.

Conclusions

The Netherlands lags behind its European commitments for renewable energy production by 2020. Despite the adoption of a new package with measures to intensify renewable energy capacity investment, according to a review by ECN (ECN, 2017) and a recent European Commission report on member states' compliance with the Renewable Energy Directive (European Commission, 2017), the country is expected to fall short of the 14% renewable energy target by 2020. Biomass is currently the main source of renewable energy in the Netherlands and wind energy deployment has recently accelerated. However, for compliance with 2020 and future targets, solar energy opportunities will also need to be utilised. For that, solar-based energy will have to catch up strongly as, for example, solar PV was only responsible for less than 1% of Dutch electricity production in 2017 (about 14% of the gross production of the electricity was generated from renewable sources, to which solar energy contributed 13%).

In order to understand uncertainties around scaling up solar energy in the Netherlands and how these could transform into risks, this chapter has focused on expanding electricity production through solar (PV). We did this by assessing the pathway of upscaling small-scale solar panel use in the built environment (e.g. on rooftops of households, small businesses, and schools) and large-scale applications of solar panels on land and infrastructure (e.g. larger-scale rooftop projects and ground-mounted solar parks).

Essentially, for reaching the 14% renewable energy target as well as targets beyond 2020, the Netherlands needs both solar PV options. At the same time, it is realised that scaling up solar PV can have negative impacts on society and the economy, which could be considered as consequential risks. Moreover, due to the existence of several implementation barriers there is a risk that a pathway cannot be implemented or will be implemented at a later stage or on a smaller scale. Both types of risks could block expanding solar PV in the Netherlands at the scale desired for complying with EU energy and climate commitments. For instance, replacing conventional energy sources with solar PV applications may have negative impacts on grid stability or could negatively affect the government's fiscal budget. Examples of implementation risks are that solar parks may not fit into existing spatial plans, may face social resistance, and that only a small fraction of dwellings and buildings are suitable for panels.

Notes

1 JIN Climate and Sustainability, Groningen, the Netherlands. We acknowledge the support by Gert-Jan Kok during the preparation of the manuscript.
2 The SDE+ subsidy scheme, called Stimulering Duurzame Energieproductie (or 'support to sustainable energy production'; the '+' refers to the revision of the scheme in 2011), is based on a feed-in tariff which guarantees that renewable energy producers receive an electricity price that is sufficient for covering investment and exploitation costs with a fair profit margin for the entire subsidy period. For each round, the feed-in tariffs are determined by the government for each renewable energy technology, for a fixed amount of full load hours and adjusted to market developments. The SDE+ budget is allocated according to a merit order.
3 With a small-scale energy connection of 3 × 80 Amp.
4 Risk elicitation of expanding solar PV in other member states can be found in Lüthi and Wüstenhagen (2012), Dusonchet and Telaretti (2015), and Del Rio and Mir-Artigues (2012).
5 This policy provides a financially attractive incentive for households and small businesses that produce solar power, use part of it, and feed the rest back to the electricity grid (net metering is then the difference between what is taken from and given to the grid).
6 The modelling exercise was done with help of the Business Strategy Assessment Model (BSAM). It is an agent-based model that is used in the TRANSrisk project for case study analysis.

References

CBS (2017). *Share of renewable energy at 5.9% in 2016.* Available at: www.cbs.nl/en-gb/ news/2017/22/share-of-renewable-energy-at-5-9-in-2016 [accessed 9 April 2018].

CBS (2018). *Hernieuwbare elektriciteit; productie en vermogen.* Available at: http://statline. cbs.nl/Statweb/publication/?DM=SLNL&PA=82610ned&D1=a&D2=5&D3=24-27&HDR=T&STB=G1,G2&VW=T [accessed 30 April 2018].

Cobouw (2018). *Pijplijn aan zonnepark-projecten zit bomvol; maar zijn er voldoende installateurs?* Available at: www.cobouw.nl/bouwbreed/nieuws/2018/05/pijplijn-aan-zonnepark-projecten-zit-bomvol-101260561?vakmedianet-approve-cookies=1&_ga=2.80199511.1330929209.1537350734-1140794003.1537350734.

Del Rio, P., Mir-Artigues, P. (2012). Support for solar PV deployment in Spain: Some policy lessons. *Renewable and Sustainable Energy Reviews* 16(8), 5557–5566.

DNV GL (2016). *Nationaal actieplan zonnestroom 2016.* DNV GL, Arnhem.

DSO (2017, 30 November). Author's interview with grid operator.

Dusonchet, L., Telaretti, E. (2015). Comparative economic analysis of support policies for solar PV in the most representative EU countries. *Renewable and Sustainable Energy Reviews* 42, 986–998.

ECN (2017). *Nationale Energieverkenning 2017.* ECN, Petten.

European Commission (2017). *Renewable Energy Progress Report.* European Commission, Brussels.

Gemeente Súdwest-Fryslân (2015). *Ruimte voor de zon.* Available at: www.topentwelonline. nl/uploads/Dorpsbelang/notitie%20Ruimte%20voor%20de%20zon%20college12-7.pdf.

Kamp, H. (2016). *Kamerbrief intensiveringspakket Energieakkoord.* Available at: www.rijks-overheid.nl/documenten/kamerstukken/2016/05/17/kamerbrief-intensiveringspakket-energieakkoord.

Lüthi, S., Wüstenhagen, R. (2012). The price of policy risk – Empirical insights from choice experiments with European photovoltaic project developers. *Energy Economics* 34(4), 1001–1011.

Nikas, A., Doukas, H., van der Gaast, W., Szendrei, K. (2018). Expert views on low-carbon transition strategies for the Dutch solar sector: a delay-based fuzzy cognitive mapping approach. *IFAC-PapersOnLine* 51(30), 715–720.

RVO (2017). *Monitor Wind op Land 2016.* Rijksdienst voor Ondernemend Nederland, Utrecht.

Van de Veen, K. (2018). *Waarschuwing: geen nieuwe zonneparken, het net kan het niet aan.* Available at: https://nos.nl/artikel/2227659-waarschuwing-geen-nieuwe-zonneparken-het-net-kan-het-niet-aan.html [accessed 17 April 2018].

Van den Eerenbeemt, M. (2018, 15 May). *Een confetti van zonneparken. De Volkskrant,* pp. 26–27.

VVD, CDA, D66, ChristenUnie (2017). *Vertrouwen in de toekomst – Regeerakkoord 2017–2021.* Rijksoverheid, Den Haag.

Zon op Kaart (2018). *Zon op kaart.* Available at: http://zonopkaart.nl/ [accessed 15 May 2018].

8 Spain

On a rollercoaster of regulatory change – risks and uncertainties associated with renewable energy transitions

Alevgul H. Sorman, Cristina Pizarro-Irizar,
Xaquín García-Muros, Mikel González-Eguino,
and Iñaki Arto

Introduction

Spain, in alignment with the European Union's (EU) climate policy, aims at reducing greenhouse gas (GHG) emissions to 80–95% below 1990 levels by 2050 (European Commission, 2011). In the Paris Agreement era, the EU's Intended Nationally Determined Contributions (INDC) sets the binding target of an at least 40% domestic reduction in GHG emissions by 2030 compared to 1990 (European Commission, 2015). Additionally, the 2030 Energy Strategy for the EU (European Commission, 2014) includes a target of at least a 27% share of renewable energy (RE) consumption – binding at the EU level – and at least 27% energy savings – non-binding – compared with the business-as-usual scenario. These targets are to be fulfilled jointly with all member states, yet there are currently no binding targets for individual member states for the post-2020 period.

Since the end of 2017, Spain has picked up its commitment to complying with the objectives set out in the Paris Agreement and within the framework of the European Union, and has begun an elaboration of the Climate Change and Energy Transition Law (Under the Article 26.2 of Law 50/1997) of the government.

Although currently Spain is well off in meeting EU targets, changes in its regulatory system over the years have caused disturbances in the roll out and ambition of renewable energy sources (RESs). These sources cover wind power, solar power from photovoltaics (solar PV), concentrated solar power (CSP), small-scale hydropower technologies (<50 MW), waste, waste treatment, and geothermal power.

Renewables, especially in the electricity sector (RES-E), during the greatest renewable expansion period in Spain (from 2004 until 2013), were supported by a combined system of feed-in tariffs (FIT) and feed-in premiums (FIP) which was established in the Renewable Energy Act and applied to all RES and cogeneration (BOE, 2004). In this period, RES-E grew from a 9% of gross generation in 2004 up to 26% in 2015 (REE, 2004–2015). However, one of the most important political concerns at that time was the tariff deficit: the gap between the cost of electricity generation and what consumers pay. Theoretically, electricity revenues should cover for costs, set by the regulator (Paz Espinosa, 2013) yet this deficit, increasing from €0.25 billion in the year 2000 to €26 billion in 2013, had become so big that it was triggering a financial black hole in the economy (CNE/CNMC, 2007–2013). The tariff deficit in Spain, gradually mounting since the 2000s, came from generous subsidies to renewables under the special regime to support to renewable and co-generation potentially linked to over-ambitious infrastructure planning, but also from growing costs in extra peninsular islands, as well as network costs (Johannesson Linden *et al.*, 2014). Solving the deficit issue through increasing consumer prices was difficult as electricity prices in Spain were already high (Paz Espinosa, 2013). In turn, subsequent regulatory changes in the hope of closing the tariff deficit have burdened the renewable energy sector due to retroactive subsidy cuts leading to great uncertainties, emerging risk factors, and numerous examples of energy injustices that will be scrutinised throughout the chapter.

Thus Spain, once a founding example of the roll out of renewables, followed an overall retreat due to retroactive regulatory and economic conditions, leaving concerns and uncertainties about future pathways within the Spanish renewable energy panorama. In fact, before the phasing out of the FIT–FIP system, Spain had ranked third in total renewable power capacity per capita (excluding hydro) by the end of 2012 (REN21, 2013) on the global scale, but by the end of 2016 it had recoiled to the fifth/sixth place (REN21, 2017).

In this chapter we explore energy transition strategies, particularly with respect to regulatory changes affecting solar energy rollout in Spain. Stakeholder insights based on past experiences are presented as 'lessons learned' focusing on perceived risks including potential cases of energy injustices. This knowledge needs to feed back into the creation of new low-carbon pathways, which we present as potential policy mixes identified per sector (transport, building, and industry).

A rollercoaster ride: a review of regulatory changes in Spain

Renewable energy (RE) in Spain has been regulated since 1980, when Law 82/1980 (BOE, 1980) on energy conservation was enacted, which claimed an increase in energy efficiency and a reduction in energy dependence. It happened as a consequence of the second international oil crisis and represented the start of the development of RES in the country. Since then, RE promotion has been a national policy priority and legislation has been in constant change. In 1985, the government firmly pledged its commitment to RE with the Royal Decree 916/1985 (BOE, 1985) supporting small-hydraulic energy, the only RES existing at that time.

After entry into the European Union, Law 54/1997 (BOE, 1997) liberalised the electricity sector in Spain and established a plan for the promotion of RES for achieving the goal of 12% of gross inland consumption of energy from renewable sources by 2010 (IDAE, 2005). After this law, the Special Regime (SR) and the corresponding FIT scheme were first established. The SR included renewable energy (wind, solar PV, solar thermal, small-scale hydropower, and biomass/wastes) and co-generation generators, with a maximum of 50 MW power.

Later, Royal Decree 436/2004 (BOE, 2004), known as the Spanish Renewable Energy Act, was set up to fit into the existing general framework supporting the electricity from RES/RES-E). Generators could decide to sell their electricity to a distributor and receive a fixed tariff (FIT) or sell it on the free market and receive a premium tariff (FIP) on top of the market price. The main difference between FIT and FIP is that the incentive level under the FIT system is fixed, whereas under the FIP scheme renewable generators get a guaranteed premium, which is lower than the FIT, plus the price of the pool. This decree provided incentives for new RES installed capacity and was subsequently renewed in the Royal Decree 661/2007 (BOE, 2007a), where new tariffs and premiums for RES-E generators, as well as a cap and a floor for renewable remuneration, were established.

The combined system of FIT and FIP led to a strong increase of investment in electricity production from renewable energy so that most technologies highly exceeded government targets for the period 2005–2010 (IDAE, 2005). In an attempt to reduce regulatory costs, incentives were adjusted in 2010 (BOE, 2010), including cuts to the FIT of solar thermal electricity and wind generation, and a cap on the number of hours eligible for support for PV installations.

In a context of overcapacity and weak demand, the regulatory changes introduced in 2010 were not deemed sufficient to reduce regulatory costs, and therefore in 2012 a new regulation (BOE, 2012) was passed for the temporary suppression of FIT and FIP for new installations. These measures left new RES-E without financial support but existing obligations remained.

The last cutbacks to the FIT–FIP system were passed in 2013 (BOE, 2013a, 2013b) and affected all RES units, including those that were already functioning. The FIT and FIP were reduced for both new and existing generation plants. This reform aimed at stabilising financial fluctuations in the RES system that had contributed to the accumulated €26-billion tariff deficit by 2013 (~2% of Spanish GDP). From 2014 onwards (BOE, 2014a, 2014b) renewable energy producers receive the market price and, if needed, a subsidy to guarantee a fixed rate of return on investment (the yield of the ten-year Spanish treasury bond plus 300 basis points), which is subject to review every six years.

The new scheme consists of the regular electricity market price supplemented by a capacity payment and a generation-based premium which are calculated on the basis of technology- and project-specific parameters. This new framework includes several remuneration adjustment mechanisms which could yield additional risks compared to the former system, based on a pre-established long-term tariff.

Another relevant issue affecting the future of RE in Spain is related to the role of RE in distributed generation. In this sense, Royal Decree 900/2015 (BOE, 2015) legislates electricity generation for self-consumption. It applies to any RES generating electricity for self-consumption *and* being connected to the national grid, which will be subjected to access and consumption fee charges. The Spanish government claims that this regulation has been designed in order to ensure the technical and economic sustainability of the national grid. However, public opinion labelled the initiative as a 'sun tax' since consumers with their own PV systems are to be taxed for the electricity they generate and consume if they are connected to the grid. This regulation in essence has not considered the external benefits that these consumers generate to the electricity system while also complying with meeting the renewables and climate objectives. Moreover, failure to comply with the law or non-registration as a self-consumer are subjected to a financial penalty between just over €6 million and €60 million, deterring small actors (IEA, 2015).

The Spanish legal framework had direct implications for the evolution of renewable energy's installed capacity and electricity generation. As a result of national regulation, when the FIT–FIP-oriented SR was in force, capacity grew from 17 GW in 2004 to almost 40 GW in 2013, and electricity generation increased from 47 TWh in 2004 to 111 TWh in 2013. However, the last energy reform (including the subsequent cutbacks in the incentive scheme from 2012 onwards) led to no new renewable energy capacity in 2014 and 2015, and a drop in the electricity generated by the SR. By technology, wind and solar accounted for 74% of total SR capacity in 2015 (57% wind, 12% solar PV, and 6% solar thermal) and 62% of electricity generation (49% wind, 8% solar PV, and 5% solar thermal).

The costs of public retribution to RES and co-generation in Spain rose steeply from 2008 to 2013 (Figure 8.1). This coincided with the period of maximum investment in renewable energy capacity and dropped after the suppression of the FIT–FIP scheme.

The regulatory framework for RE in the transport sector sets a biofuel quota of 4.1% of the total amount sold from 2013 onwards (BOE, 2013c: Article 41). This target represents a reduction from the original 6.5%, thus endangering the achievement of the 10% goal by 2020 (risk) but at the same time incurs some additional benefits (e.g. reduces the biofuels competing for food production or avoids investment in alternative energy sources which do not generate high returns on the energy invested, having low energy return on investment (EROI) (Cleveland *et al.*, 2016).

Regarding the heating and cooling sector, European directives have not been properly transposed into the Spanish legal framework, thereby limiting the incentives to low-carbon sources for heating and cooling (H&C). The only policies aimed at RES-H&C include the building sector, where the Regulation for Thermal Installations in Buildings encourages the use of RES (biomass, geothermal, and solar) with energy efficiency purposes, but with no concrete targets (BOE, 2007b; updated in BOE, 2013a).

Finally, concerning the energy efficiency target, the Spanish government forwarded the National Energy Efficiency Action Plan (NEEAP) 2014–2020 to the European Commission. This strategic plan contains targets in line with the

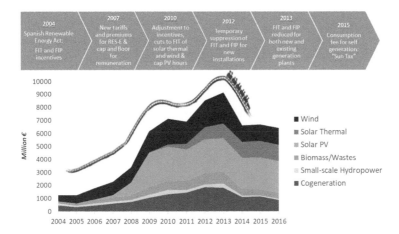

Figure 8.1 Timeline of regulatory changes and incentive schemes' variation for RES and co-generation in Spain.

objective of improving energy efficiency by 20% by 2020 (MINETUR, 2014) with the adoption of an energy efficiency scheme, financial and fiscal measures, as well as energy efficiency standards and voluntary agreements.

Research process and methods

In line with the debate in Spain, especially relevant now with a deliberative process on the design of the Law of Climate Change and Energy Transition, the authors have carried out a combination of: (i) a comprehensive review of past and existing regulations on renewable energy in Spain; and (ii) an elaborate expert stakeholder engagement process. The stakeholder engagement process in itself has been conducted initially through: (1) an in-depth interview process; and thereafter (2) a far-reaching survey to provide insights for the renewable energy transition, as detailed later. The narrative told in this chapter is the result from all these efforts. Previous policy analyses and regulatory changes are compiled to deduce risks and negative consequences which are then used for inferring several implementation risks. These are resumed under occurring energy (in)justices in order to serve as lessons learned in the design of the new Climate Change and Energy Transition Law of Spain.

1 In-depth interviews with expert stakeholders

Out of an initial outreach to 24 expert stakeholders in the field, 16 experts were interviewed, either in person or via Skype, lasting approximately 45–60 minutes. Out of the 16 interviewees, the authors had a sample distribution across sectors of 30% energy generators, 13% government/administration, 13%

NGO representatives, and 44% academic representatives. Of the 16 interviewees, four were females (25%).

The semi-structured interviews were conducted in two blocks, the first block addressing controlled questions on the diagnostic of the energy panorama in Spain, regarding decarbonisation and the promotion of renewables in Spain in the past and current times. Thereafter, stakeholders' narratives on their visions for the future (2020–2030–2050) were asked based on the energy mix, instruments and infrastructures, criteria, actors, and limitations, along with positive and negative consequences.

2 Survey

Based on the answers compiled from the 16 responses and over 15 hours of interviews, the authors put together a 23-question survey, four of which are covered within the scope of this chapter detailing questions on policy mixes for future transition pathways. The 15-minutes overall survey reached 206 respondents. The four questions tackled here provide insights on policy mixes for future transition pathways that stakeholders identified as crucial elements to be considered. More specifically, these questions grasp the diversity of perceptions regarding the importance and probability of certain policy measures in the transportation, building, and industrial sector. The 'importance' question as posed (answering from a dropdown menu of 'low'/'medium'/'high'/'not applicable' options) revealed how stakeholders weighed each policy mix in terms of its relevance for its value, magnitude, and influence, whereas the probability criteria (once again from a selection of 'low'/'medium'/'high'/'not applicable') assessed the possibility that this policy was likely to happen.

The extent of responses, reaching 206 stakeholders, resulted in a split with 63% replies coming from men and 37% from women. In terms of response rates, there was also an inclined academic bias (40% of responses) in the sampling of sectors of activity, with energy suppliers and the private sector constituting 27%, NGOs 8%, administration 16%, and others 9% of the sample.

The mixture of these methods has a two-fold objective: initially, to draw lessons from the past in order to build sustainable pathways for a low-carbon future in Spain, based on risks perceived and emerging justice concerns; and thereafter to incorporate a multitude of different ontologies and stakeholders, integrating the vision of regulators, academics, NGOs, and members of the private sector.

Reframing regulatory changes as uncertainties, risks, and justice concerns

From the roller coaster trajectory of past and existing regulation on renewable energy in Spain lessons emerge: *negative consequences* resulting from policy choices in the rollout of renewable energies and *implementations risks* of low-carbon policy pathways that need to be considered for future prospects for the design of the new Climate Change and Energy Transition Law of Spain (Figure 8.2).

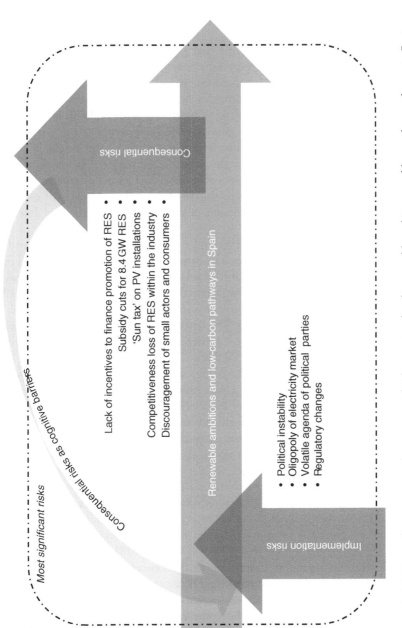

Consequential risks

- Lack of incentives to finance promotion of RES
- Subsidy cuts for 8.4 GW RES
- 'Sun tax' on PV installations
- Competitiveness loss of RES within the industry
- Discouragement of small actors and consumers

Renewable ambitions and low-carbon pathways in Spain

- Political instability
- Oligopoly of electricity market
- Volatile agenda of political parties
- Regulatory changes

Implementation risks

Consequential risks as cognitive barriers

Most significant risks

Figure 8.2 Main implementation and consequential risks associated with renewable ambitions and low-carbon pathways in Spain.

Consequential risks and the energy transition panorama

As a result of the political panorama and regulatory shifts, several consequential risks have been observed grounded on the archival research on regulatory changes and stakeholders' interviews.

Foremost, it has been underlined that gender representation has been a reoccurring debate in Spain, especially relevant within the energy sector which has the lowest percentage of women with executive positions (Pena, 2016). This has come about in terms of negative consequences including a gender-plural vision in decision-making mechanisms in building low-carbon energy transition pathways. The gender under-representation of our stakeholder interviews and survey results, within the scope of this research, have also indicated a representation of the snapshot of reality. The peak of the debate, and hence an imminent consequential risk of a future decarbonisation pathway, has also been observed in the 14-strong all-male panel of experts that have been selected in the design of the Climate Change and Energy Transition Law of Spain.

Regarding regulatory changes, stakeholder interviews have helped to identify numerous major consequential risks. The Royal Decree of 2013 implemented by the state foresaw retroactive cuts of RES subsidies in terms of protective measures so as not to increase the aforementioned tariff deficit. These cuts accounted for 8.4 GW or 37% of the cumulative investment, around €1.7 billion in subsidies for the sector (Rincón, 2016). The renewable energy sector, being stripped of all subsidies, faced a major consequence of the Spanish state breaking its promise to maintain incentives for 20 years.

Royal Decree 900/2015 (BOE, 2015a) resulted in the taxation of self-consumption for PV installations. Under this decree, Type 1 modality of prosumers indicates that if solar panel installations of less than 100 kW generate excess electricity, then that excess amount has to be donated to the grid and no remuneration is given. PV prosumers exceeding 100 kW (Type 2), regardless of whether they are using the grid or not at that moment of time, are subject to two types of taxes. The changes in the regulatory system oblige a self-consumption PV owner to pay a charge for the whole power (in terms of kW capacity) installed (the power contracted through the electricity company *plus* the power from the personal PV installation) as well as a (second) 'sun tax' for the *electricity generated for self-consumption* from the personal PV installation (applied to installations larger than 10 kW) (Tsagas, 2015).

Evidently, this regulatory change, rather than encouraging the participation of small actors and consumers that would promote decentralisation as a means of achieving energy and climate action led by community-owned, renewable energy systems, has resulted in the contrary (Angel, 2016), penalising self-consumption and generation tremendously. Not only does this portray an unfeasible (both economically and socially) option for households, but it may also result in consequential risks for businesses: industries are hindered in PV installation, affecting the overall competitiveness of an RES transition within the industry, with the lack of incentives to finance the promotion of renewables, leading to competitiveness losses.

Along similar lines, the increase of the price of electricity to end users has especially been affecting the more vulnerable households and industries. RES support is currently an important component of the regulated cost of electricity. Thus, due to the need to finance relevant RES subsidies by means of an electricity price surcharge and the lack of competitiveness in the electricity sector, electricity prices in Spain have increased significantly. Despite being connected to mainland Europe, interconnection is limited. In fact, Spain is commonly known as an 'energy island' and therefore one of the European plans for the 2030 roadmap is to increase interconnections in this sense. Spaniards currently pay the third-highest electricity tariffs in Europe, after Cyprus and Malta – island countries highly dependent on imported oil products for electricity generation. As such, regressive effects of regulations are most often impacting those who are least able to afford it, who are forced to bear the costs while benefiting from relatively few advantages. The poorest households are even more affected by higher electricity prices since they expend a greater proportion of their income on electricity. Spending on electricity as a proportion of disposable income in the poorest households is around 5%, whereas in the richest households it is around 1% (Garcia-Muros, Böhringer, and Gonzalez-Eguino, 2017). These regressive impacts can reduce the political feasibility of new measures to promote renewable energy sources. Public acceptability is essential for effective mitigation policies to be adopted, and equity and fairness play an important role in how such measures are regarded by public opinion (Bristow *et al.*, 2010).

Despite the social bonus mechanism (BOE, 2009) – presenting a 25% discount on the total electricity bill for eligible households – for protecting vulnerable consumers with low incomes, it has been observed that energy poverty in Spain increased from 3.6% before the economic crisis in 2007 to 9.88% in 2013 (nominal values) after the average electricity bill grew by 76% and natural gas bills by 35% (Economics for Energy, 2015).

Implementation risks and the roll out of renewables

Stakeholders have portrayed an evident level of pessimism due to cutbacks in the renewables roll out due to regulatory changes, especially from 2013 onwards. Uncertainty lingers along with the highly volatile agenda of political parties, oftentimes changing based on election cycles. Moreover, additional compensation adjustment mechanisms along the timeline add overall uncertainty and instability to investments in the system, potentially aggravating additional risk factors in the future.

One of the biggest and foremost characteristics within the Spanish electricity sector is that over 75% of electricity generation and over 85% of sales (Unesa, 2013) are controlled by five major companies (Iberdrola, Gas Natural Fenosa, Endesa, Viesgo, and EDP), meaning that the electricity market presents a typical structure of an oligopoly. In this type of market structure the risk of lack of competition increases, along with the risks of lack of independence of regulatory bodies due to the so-called 'regulatory capture' process. Such a practice undermines the interests of consumers and also creates the implementation risk of policies aimed at promoting public interest, such as those related to a boost for a renewable energy transition.

Moreover, other factors aggregating to implementation risks arise from different parties in the government having different positions about the role of renewables and how they should be promoted and financed. The regulatory rollercoaster ride, as aforementioned, results in RES developments being legally and institutionally hindered and results in regulatory barriers continuously changing throughout the years (Capellán-Perez, Campos-Celador, and Terés-Zubiaga, 2016).

Bridging risks to justice concerns: the energy justice framework

These illustrated risk factors can be brought together neatly within the field of *energy justice*, which calls for 'a global energy system that fairly disseminates both the benefits and costs of energy services and one that has representative and impartial energy decision-making' (Sovacool and Dworkin, 2015). Such an approach is translated to the operationalisation of energy justice theory across four primary axes: *recognition, procedural, distributional*, and *restorative* components.

It is accepted that recognition justice embraces the multitude of actors, voices, and positions, while procedural justice assumes inclusive, transparent decision-making mechanisms as well as dissolved power structures embedded in energy governance (Sovacool and Dworkin, 2015). Distributional justice implies responsibility across nodes of production and modes of consumption (Walker, 2012) and assumes fair distribution among producers and consumers, while restorative justice highlights the need for repairing the harm done to people (and/or society/nature) and assisting in pinpointing where prevention needs to occur (Heffron and McCauley, 2017).

Energy injustices play out through technological, social, political, institutional, or spatial ways (Bridge *et al.*, 2013; Sovacool and Dworkin, 2015; McCauley, 2018). Thus, with emerging consequential risk factors in the RES sector and recent policy changes in Spain, it is seen that energy injustices have been emerging along the four components, some of which are highlighted in the following paragraphs. These are to serve as illustrative examples for providing a series of 'lessons learned' from cases, for deriving alternative narratives for change, and for ensuring energy justice application in policy and practice in the future.

Initially, the oligopolistic structure of the electricity system in Spain can be presented as a concern in terms of *procedural injustices* in relation to the European Commission Directive 2009/72/EC, regarding the functioning of the internal market of electricity. The directive, aimed at ensuring an equal level playing field in generation, has been contested in Spain by failing to ensure effective separation between the companies involved in the generation, distribution, and sale of electricity. The same market has also contributed to the creation of a tariff deficit resulting from the difference between the income of the electricity system and its regulatory costs. The tariff deficit reached €26 billion in 2013 (CNE/CNMC, 2007–2013), burdening both the RES sector and end users. Thus, the energy policymaking processes have been criticised for being developed unfairly by powerful actors while excluding the voices of those potentially or actually affected (Clayton, 2000).

This can illustratively be seen in Spain, in this sense a unique case being the only country where self-consumers are taxed for the electricity generated for their own on-site usage beyond a normal fee for electricity exported to the grid as in most counties. A *Manifesto for the repeal of The Royal Decree 900/2015* has been developed contrary to these procedural injustices, arguing that self-production and consumption constitute a community right and that systems based on decentralised RES increase energy efficiency, create jobs, and boost the local economy (Afman *et al.*, 2017).

Second, the retroactive cuts of RES subsidies, as a response to the tariff deficit (the Royal Decree of 2013) that broke the Spanish state's promise, is a foremost violation of *recognition* as the decree has not had any prior negotiation with stakeholders (including RES investors), as highlighted repetitively by the national wind association, AEE (McGovern, 2014).

As a result, in June 2016 hundreds of cases were brought before the Supreme Court by investors who were against the legislative changes in an attempt for *restorative* claims, yet the court ruled against the investors. The legal battle in terms of retroactive cuts is not over yet. Now, many such investor companies have turned to the International Centre for Settlement of Investment Disputes with claims against Spain, with results still pending. Most recently, on 23 February 2018, in a second international ruling against retroactive cuts, Spain was coerced to pay a Luxembourg-based investment firm €53 million in compensation (Weyndling, 2018).

Another very important dimension to consider in the renewable roll out is the uneven allocation of risks and benefits of spatial factors and in terms of *distributional* questions. An extension and expansion of new geospatial requirements based on a shift to low-carbon alternatives (Bridge *et al.*, 2013; Bouzarovski and Simcock, 2017; Scheidel and Sorman, 2012), whether for solar or wind, come about with distributional implications across both those regions producing this energy and those consuming it. A case in the southern Catalonia region has been well documented, where despite providing a surplus of electricity through wind production the region is still bound to the political and economic periphery (Franquesa, 2018).

Distributional implications also resulted through end users with disproportionate access to energy services and impacts on wellbeing in terms of energy poverty. Concerns about energy poverty have been on the rise with estimates indicating more than 1.5 million Spanish households in a situation of energy poverty, meaning that over five million people have serious difficulties in meeting basic energy needs such as electricity and gas (ACA, 2016; Raso,

2017). Such distributional injustices affect the wellbeing of those appearing to be most vulnerable, often mapped out into low-income households.

Future pathways and stakeholders' ambitions

The regulatory ambiguity and arising concerns in terms of energy injustices open up new potential take-home messages in the design phase of the Climate Change and Energy Transition Law. The understandings emerging from embedded risk factors and uncertainties have come about as lessons learned in the past, opening up the space to follow alternative policy mixes as identified by stakeholders, and laying the foundations of a future plausible low-carbon energy pathway for Spain.

The key question that remains is whether Spain should carry out more ambitious decarbonisation processes than those set by European directives. The results of our stakeholder survey composed of academics, administration, NGOs, energy providers and the private sector, and others have identified possible outcomes in terms of the impending energy transition that awaits Spain.

Several pathways towards the future exist. More conservative and cautious objectives are laid forth, on setting the 27% share of renewable energy (RE) consumption – binding at the EU level – as indicated by 9% of stakeholders participating to the survey. Yet a majority of stakeholders (53%) see the energy transition pathway and ambitious renewable objectives as a great opportunity for Spain. This provides an opportunity not only to pursue low-carbon alternatives but also to engage all important stakeholders in the field in the design of such ambitious policies. The middle grounds in terms of such an ambitious transition are supported by 38% of the engaged stakeholders, who believe that such a transition would be good for Spain as long as the additional costs that may arise in such a process are also considered and internalised.

Narratives depicted by each individual stakeholder group regarding their ambitions in future pathway options remain diverse yet ambitious overall (Figure 8.3). While NGOs are on the frontlines of determined decarbonisation

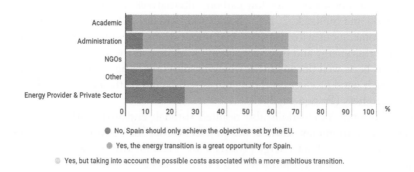

Figure 8.3 Pathways for Spanish energy transition as narrated by stakeholder groups.

policies, more cautious outlooks are portrayed by energy providers and the private sector.

Policy mixes for sectoral transitions in Spain

With the attempt of identifying potential low-carbon transition pathways within each distinct sector, stakeholders have categorised policy proposals according to their importance – defined as their relevance for its value, magnitude, and influence – and probability – defined as the likelihood of this measure happening (Figure 8.4). Thus the authors have embraced a two-fold objective with such an approach: initially, defining entry points of sectoral intervention of policies where stakeholders believe there is an impact for change; and second, to measure the likelihood of carrying out such a policy, identifying whether there are barriers to its implementation and thus risks or prospects for such a policy being carried out. The multitude of criteria are to help grasp how different narratives have emerged as alternative policy proposals when tackling the renewable energy transition ambitions for Spain overall.

Primarily, most policy mixes portray high stakes and a high likelihood of occurrence. Indeed, none were identified within the low probability and importance quadrant. Although most are valued highly important, some are considered less likely to happen. A greater reuse of materials and processes within the industrial sector; and efficiency improvements in rehabilitation in buildings rank as of utmost importance, followed by policies on reduction and changes in

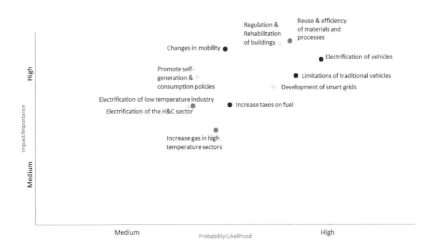

Figure 8.4 Policy mixes by sector, positioned according to importance and probability of occurrence.

Note
Results are plotted with portability/likelihood along the horizontal, the importance/impact across the vertical, divided by colour for different sectoral policy mixes.

mobility through public transport, railroad, and car sharing, though transport policies have resulted in highest prominence. Striking as it may seem, electrification of vehicles and limitation of use and access of traditional vehicles in cities seem more likely to take place.

In the industry, indicating policies for greater reuse of materials and processes within the sector rank with utmost impact, as recycling and upcycling are becoming integral actors in production chains. Such measures within the industry push forth a triple objective in terms of energy efficiency, the zero-waste initiative, and a circular economy. Electrification of low-temperature industry (for example the textile sector and food processing) is seen as of medium importance but lesser possibility, while increases in the use of gas in high-temperature sectors ranked with least favourable potential due to prolonging fossil-fuel reliance, though more likely to happen.

Transport has been identified as a key area of action by stakeholders as electric vehicles (EV) slowly infiltrate the car market, while incentivising measures on reduction and changes in mobility through public transport, railroads, and car sharing are classified as having a major impact. Penetration of EVs into policy arenas is seen as highly probable, and is supported by big car industries (e.g. Volvo) taking measures to either partially or completely switch to battery-powered engines, ending traditional internal combustion engine vehicles (Vaughan, 2017). Limitations to the use and access of traditional vehicles into cities are also seen as potential policies highly likely to be implemented. Several cities are starting to implement policies in this direction. For instance, Barcelona is a successful case: since the initial probation of the 'superblocks' policy (limiting the interior spaces of a 9×9 series of blocks to urban mobility, increasing pacified spaces, and giving priority of pedestrians) as a part of its urban mobility plans, Barcelona has shown successful transition strategies (Ajuntament de Barcelona, 2014). Madrid has also planned to ban the oldest and most polluting vehicles, prohibiting the use of gasoline cars registered before 2000 and diesel-powered cars registered before 2006 (Plan A Madrid, 2017). In the survey, increasing taxes on transport fuels resulted emerged with a medium-high probability of occurrence but less important compared to other transport policies identified, despite the fact that the International Energy Agency (IEA) has pointed out that oil taxation in Spain is quite low; e.g. the tax component on the total diesel price is only 51%, whereas in the United Kingdom it is 67% and in Italy it is 62% (IEA, 2015).

Improvements in energy efficiency and building restoration has ranked second in importance as currently there are no legislative developments sustaining operations of urban rehabilitation, regeneration, and renewal, and legal obstacles prevail preventing their implementation. Thus, legislative amendments making arrangements more flexible, opening up new mechanisms for funding, and encouraging public–private partnership are on the agenda of the Spanish Building Assessment Report (IEE), in addition to including a Certification of Energy Efficiency within the scheme. Policies regarding promoting self-generation and consumption rank of high importance, yet reveal medium probability of occurrence. A major reason behind this may be the regulatory changes de-incentivising and furthermore penalising RES with additional taxes (e.g.: the aforementioned 'sun tax'). Though not as important, promoting the development and promotion of smart grids for control and monitoring of electricity consumption is more likely to happen. While the promotion of the electrification of the heating and cooling (H&C) sector is perceived to be important, it features medium possibility of taking place according to stakeholders.

Policies supported by technological advancements or change such as the smart grid, electric vehicles, or industrial improvements in production processes are more likely to happen, rather than policies with higher impact that require regulatory and behavioural changes.

These policy mixes are to serve as an initial diagnostic where a great range of stakeholders believe it is necessary to act in an attempt to meet ambitious renewable energy transition targets.

Conclusions

In Spain, the tariff deficit increased over ten times within 13 years (from €0.25 billion in the year 2000 to €26 billion in 2013). However, regulatory changes in the false hope of closing this tariff deficit have burdened the progress of renewables in the energy sector. Regulatory frameworks, including the Royal Decree on self-consumption and retroactive subsidy cuts, have hindered RES production and development in Spain and also put future investments in jeopardy, leading to great uncertainties and risks. Moreover, the interplay of these factors has played out through numerous procedural, recognition, distributional, and restorative energy injustices that have been highlighted within the scope of this chapter. Such past developments, presented as 'lessons learned' aimed at renewable energies, should, for example, be considered in the future of policy design in other support schemes targeting energy efficiency improvement, development of high-efficient co-generation (CHP), or district heating (DH) municipal systems.

The results of our stakeholder engagement and survey responses indicate that three potential pathways for an energy transition await in Spain. Less likely is to target more conservative policies aimed at reaching a 27% share of renewable energy (RE) consumption – binding at the EU level. Second is to set ambitious targets while assuming additional costs that may arise upon adopting such a low-carbon transition process. However, the majority of stakeholders indicate the energy transition pathway and ambitious renewable objectives as a great opportunity for Spain. The narratives of each stakeholder group, composed of academics, administration, NGOs, energy providers, the private sector, and others, have underlined the importance of aspiring for decarbonisation policies. An outlook towards particular policy instruments (covering the transport, building, and industrial sectors), based on their impact and probability of implementation in the hope of creating alternative spaces, have yielded potential pathways in enacting energy transition strategies in Spain.

Moreover, since June 2018, there have been changes on the horizon in Spain. A new Ministry for Ecological Transition, the first of its kind, has emerged and Spain now opts for more ambitious targets, and has started debating the future of fossil-fuel installations as well as potential discussions on the elimination of the 'sun tax'. Many opportunities are present for designing a deliberative and progressive Climate Change and Energy Transition Law in the hope of alternative narratives and measures for change.

References

Afman, M., Cherif, S., Rambelli G., Sina, S. (2017). *EU Energy Market Policy: Local and Regional Experience and Policy Recommendations*. doi:10.2863/11310.

Ajuntament de Barcelona (2014). *Urban Mobility Plan of Barcelona. PMU 2013–2018*. Available at: www.bcnecologia.net/sites/default/files/proyectos/pmu_angles.pdf.

Angel, J. (2016). *Strategies of energy democracy*.Rosa Luxemburg Stiftung, Brussels.

ACA (Asociación de Ciencias Ambientales). (2016). *3rd Energy Poverty Study in Spain – New Approaches to Analysis. 3er Estudio Pobreza Energética en España – Nuevos Enfoques de Análisis*. Available at: www.cienciasambientales.org.es/noticias/567-3er-estudio-pobreza-energetica-en-espana-nuevos-enfoques-de-analisis.html (in Spanish).

BOE (1980). *Law 82/1980 of December 30 (BOE 27.01.1981), N. 23: 1863–1866*. Available at: www.boe.es/boe/dias/1981/01/27/pdfs/A01863-01866.pdf (in Spanish).

BOE (1985). *Law 916/1985 of May 25 (BOE 22.06.1985), N. 149: 19398–19400*. Available at: www.boe.es/boe/dias/1985/06/22/pdfs/A19398-19400.pdf (in Spanish).

BOE (1997). *Law 54/1997 of November 28 (BOE 28.11.1997), N. 285: 35097–35126*. Available at: www.boe.es/boe/dias/1997/11/28/pdfs/A35097-35126.pdf (in Spanish).

BOE (2004). *Royal Decree 436/2004 of March 12 (BOE 27.03.2004), N. 75: 13217–13238*. Available at: www.boe.es/boe/dias/2004/03/27/pdfs/A13217-13238.pdf (in Spanish).

BOE (2007a). *Royal Decree 661/2007 of May 25 (BOE 26.05.2007), N. 126: 22846–22886*. Available at: www.boe.es/boe/dias/2007/05/26/pdfs/A22846-22886.pdf (in Spanish).

BOE (2007b). *Royal Decree 1027/2007 of July 20 (BOE 29.08.2007), N. 207: 35931–35984*. Available at: www.boe.es/boe/dias/2007/08/29/pdfs/A35931-35984.pdf (in Spanish).

BOE (2009). *Royal Decree 6/2009 of April 30 (BOE 07.05.2009), N. 111: 39404–39419*. Available at: www.boe.es/boe/dias/2009/05/07/pdfs/BOE-A-2009-7581.pdf (in Spanish).

BOE (2010). *Royal Decree 1614/2010 of December 7 (BOE 08.12.2010), N. 298: 101853–101859*. Available at: www.boe.es/boe/dias/2010/12/08/pdfs/BOE-A-2010-18915.pdf (in Spanish).

BOE (2012). *Royal Decree 1/2012 of January 27 (BOE 28.01.2012), N. 24: 8068-8072*. Available at: www.boe.es/boe/dias/2012/01/28/pdfs/BOE-A-2012-1310.pdf (in Spanish).

BOE (2013a). *Royal Decree-law 2/2013 of February 1 (BOE 02.02.2013), N. 29: 9072–9077*. Available at: www.boe.es/boe/dias/2013/02/02/pdfs/BOE-A-2013-1117.pdf (in Spanish).

BOE (2013b). *Royal Decree-law 9/2013 of July 12 (BOE 13.07.2013), N. 167: 52106–52147*. Available at: www.boe.es/boe/dias/2013/07/13/pdfs/BOE-A-2013-7705.pdf (in Spanish).

BOE (2013c). *Law 11/2013 of July 26 (BOE 27.07.2013), N. 179: 54984–55039*. Available in www.boe.es/boe/dias/2013/07/27/pdfs/BOE-A-2013-8187.pdf (in Spanish).

BOE (2014a). *Royal Decree 413/2014 of June 6 (BOE 10.06.2014), N. 140: 43876–43978*. Available at: www.boe.es/boe/dias/2014/06/10/pdfs/BOE-A-2014-6123.pdf (in Spanish).

BOE (2014b). *Order IET/1045/2014 of June 16 (BOE 20.06.2014), N. 150: 46430–48190.* Available at: www.boe.es/boe/dias/2014/06/20/pdfs/BOE-A-2014-6495.pdf (in Spanish).

BOE (2015). *Royal Decree 900/2015 of October 9 (BOE 10.10.2015), N. 243: 94874–94917.* Available at: www.boe.es/boe/dias/2015/10/10/pdfs/BOE-A-2015-10927.pdf (in Spanish).

Bouzarovski, S., Simcock, N. (2017). Spatializing energy justice. *Energy Policy* 107, 640–648.

Bridge, G., Bouzarovski, S., Bradshaw, M., Eyre, N. (2013). Geographies of energy transition: Space, place and the low-carbon economy. *Energy Policy* 53, 331–340.

Bristow, A.L., Wardman, M., Zanni, A.M., Chintakayala, P.K. (2010). Public acceptability of personal carbon trading and carbon tax. *Ecological Economics* 69(9), 1824–1837.

Capellán-Pérez, I., Campos-Celador, Á., Terés-Zubiaga, J. (2016). *Assessment of the potential of Renewable Energy Sources Cooperatives (RESCoops) in Spain towards Sustainable Degrowth.* Conference paper at the 5th International Degrowth Conference, Budapest.

Clayton, S. (2000). New ways of thinking about environmentalism: Models of justice in the environmental debate. *Journal of Social Issues* 56(3), 459–474.

Cleveland, C., Hall, C.A.S., Herendeen, R. (2006). Energy returns on ethanol production. *Science* (letters) 312, 1746.

CNE/CNMC (National Energy Commission/National Commission of Markets and Competence) (2007–2013). *Liquidaciones de actividades reguladas del sector eléctrico.* Available at: www.cnmc.es/ambitos-de-actuacion/energia/liquidaciones-y-regimen-economico (in Spanish).

Economics for Energy (2015). *Pobreza Energética en España: Análisis Económico y Propuestas de Actuación.* Available at: http://eforenergy.org/actividades/Presentacion-del-Informe-Anual-de-2014-de-Economics-for-Energy-Pobreza-Energetica-en-Espana-Analisis-Economico-y-Propuestas-de-Actuacion.php (in Spanish).

European Commission (2011). *Energy Roadmap 2050.* COM(2011) 885 final. Available at: http://eur-lex.europa.eu/legal-content/EN/TXT/PDF/?uri=CELEX:52011DC0885&from=EN.

European Commission (2014). *A policy framework for climate and energy in the period from 2020 to 2030.* COM(2014) 15 final. Available at: http://eur-lex.europa.eu/legal-content/EN/TXT/PDF/?uri=CELEX:52014DC0015&from=EN.

European Commission (2015). *Intended Nationally Determined Contribution of the EU and its Member States.* Available at: www4.unfccc.int/submissions/INDC/Published%20Documents/Latvia/1/LV-03-06-EU%20INDC.pdf.

Franquesa, J. (2018). *POWER STRUGGLES: Dignity, Value, and the Renewable Energy Frontier in Spain.* Indiana University Press.

Garcia-Muros, X., Böhringer, C., Gonzalez-Eguino, M. (2017). *Costeffectiveness and incidence of alternative mechanisms for financing renewables. BC3 Working Paper Series 2017–04.* Basque Centre for Climate Change (BC3), Leioa, Spain.

Heffron, R.J., McCauley, D. (2017). The concept of energy justice across the disciplines. *Energy Policy* 105, 658–667.

IDEA (Institute for Diversification and Energy Saving) (2005). *Plan de Energías Renovables en España 2005–2010.* Available at: www.idae.es/uploads/documentos/documentos_PER_2005-2010_8_de_gosto-2005_Completo.(modificacionpag_63)_Copia_2_301254a0.pdf (in Spanish).

IEA (International Energy Agency) (2015). *Energy Policies of IEA Countries: Spain2015 Review.* OECD (Organisation for Economic Co-operation and Development)/IEA, Paris.

Johannesson Linden, A., Kalantzis, F., Maincent, E., Pienkowski J. (2014). Electricity Tariff Deficit: Temporary or Permanent Problem in the EU? *European Commission Economic Papers* 534, October 2014, ISSN 1725–3187.

McCauley, D. (2018). *Energy Justice: Re-balancing the Trilemma of Security, Poverty and Climate Change*. Palgrave, Basingstoke.

McGovern, M. (2014). Spain passes retroactive subsidy cut law. *Wind Power*. Available at: www.windpowermonthly.com/article/1298015/spain-passes-retroactive-subsidy-cut-law.

MINETUR (Ministry of Industry, Energy and Tourism) (2014). *2014–2020 National Energy Efficiency Action Plan*. Available at: https://ec.europa.eu/energy/sites/ener/files/documents/2014_neeap_en_spain.pdf.

Paz Espinosa, M. (2013). *Understanding Tariff Deficit and its Challenges*. EFO – Spanish Economic and Financial Outlook. DFAE-II WP Series (1), 1.

Pena, A. (2016). Spain advances in the incorporation of women in the energy sector. España avanza en la incorporación de la mujer al sector energético. *El periódico de la energía*. Available at: https://elperiodicodelaenergia.com/espana-avanza-en-la-incorporacion-de-la-mujer-en-el-sector-energetico/ (in Spanish).

Plan A Madrid (2017). *Air Quality and Climate Change Plan of the City of Madrid. Plan de Calidad del Aire y Cambio Climático de la Ciudad de Madrid*. Available at: www.madrid.es/UnidadesDescentralizadas/Sostenibilidad/CalidadAire/Ficheros/PlanAireyCC_092017.pdf (in Spanish).

Raso, C. (2017). Energy poverty reaches 1.5 million households in Spain. La pobreza energética alcanza a 1,5 millones de hogares en España. *El Economista*. Available at: www.eleconomista.es/energia/noticias/8320317/04/17/La-pobreza-energetica-alcanza-a-15-millones-de-hogares-en-Espana.html (in Spanish).

REE (Red Eléctrica de España) (2004–2015). *El Sistema eléctrico español. Annual report*. Available at: www.ree.es/es/publicaciones (in Spanish).

REN21 (2013). *Renewables 2013 Global Status Report*. REN21 Secretariat, Paris. Available at: www.ren21.net/Portals/0/documents/Resources/GSR/2013/GSR2013_lowres.pdf.

REN21 (2017). *Renewables 2017 Global Status Report*. REN21 Secretariat, Paris. Available at: www.ren21.net/wp-content/uploads/2017/06/17-8399_GSR_2017_Full_Report_0621_Opt.pdf.

Rincón, R. (2016). Spain's Supreme Court backs renewable energy cuts. *El Pais*. Available at: https://elpais.com/elpais/2016/06/02/inenglish/1464860925_523010.html.

Scheidel, A., Sorman, A.H. (2012). Energy transitions and the global land rush: Ultimate drivers and persistent consequences. *Global Environmental Change* 22(3), 588–595.

Sovacool, B.K., Dworkin, M.H. (2015). Energy justice: Conceptual insights and practical applications. *Applied Energy* 142, 435–444.

Tsagas, I. (2015). Spain Approves 'Sun Tax,' Discriminates Against Solar PV. *Renewable Energy World*. Available at: www.renewableenergyworld.com/articles/2015/10/spain-approves-sun-tax-discriminates-against-solar-pv.html.

Unesa (Asociación Española de la Industria Eléctrica) (2013). Contribution of Unesa companies to the development of the Spanish society. Contribución de las compañías que integran Unesa al desarrollo de la sociedad Española. Available at: www.unesa.es/biblioteca/category/1-estudios?download=132:qcontribucion-de-las-companias-que-integran-unesa-al-desarrollo-de-la-sociedad-espanolaq (in Spanish).

Vaughan, A. (2017). All Volvo cars to be electric or hybrid from 2019. *Guardian*. Available at: www.theguardian.com/business/2017/jul/05/volvo-cars-electric-hybrid-2019.

Walker, G. (2012). *Environmental justice: concepts, evidence and politics*. Routledge, London.

Weyndling, R. (2018). Spain to pay compensation to renewables investors. *Wind Power*. Available at: www.windpowermonthly.com/article/1457902/spain-pay-compensation-renewables-investors.

9 Switzerland

Risks associated with implementing a national energy strategy

Oscar van Vliet

Introduction

In the wake of the disaster in Fukushima, the Swiss government decided in November 2011 that existing nuclear reactors would be used until no longer serviceable and then replaced by renewable sources. One plant has since been shut down and the other three, with a combined capacity of around 3 GW, are to be closed by 2034.

Switzerland already has an almost carbon-free electricity supply because of its many hydropower plants. Hydropower and other sources currently supply some 70% of Swiss electricity, which leaves around 30% of Swiss production to be shifted from nuclear to renewables. This turnaround is part of the latest Swiss Energy Strategy 2050 (ES2050). In addition to a climate target and a nuclear phase out, Switzerland intends to reduce its long-standing reliance on foreign fossil fuels, mostly oil for transportation and heating, as well as some natural gas.

Due to the unique Swiss system of direct democracy, any future expansion of wind, solar, and hydropower requires that utilities, local and regional governments, NGOs, and companies in the renewable sector work together (see *Swiss direct democracy* box and Figure 9.1). Crucially, many of the utilities are publicly and domestically owned. As a result, large amounts of political discussion and academic research have been carried out for the Swiss renewables transition over the last five years, and the process of defining and implementing the ES2050 is still ongoing.

Swiss direct democracy

The Swiss decision-making system is unique in that Switzerland is a confederation with direct democracy. The confederation aspect – the country code CH stands for its Latin name of Confoederatio Helvetica – is reflected in a very devolved government. The municipalities and cantons (analogous to provinces, counties, or states in other countries) retain a great deal of autonomy. For example, public holidays in Switzerland differ between cantons.

This autonomy stretches into the electricity sector in two ways: first, the cantons have considerable discretion when implementing the federal energy law;

and second, most of the utilities operating in the Swiss market are majority-owned by municipalities and cantons.

The direct democracy is exemplified in the Swiss system of *Volksinitiativen* ('People's Initiatives'). With a limited number of signatures (100,000 for a national initiative), any citizen can request a municipal, cantonal, or federal government to consider a specific action. If the government will not implement it out of hand, they can call a referendum on municipal, cantonal, or federal level to change policy. If a referendum passes, it becomes law in the same way as a constitutional change – parliament and the executive have no legal way of overthrowing the results.

The political landscape in Switzerland is strongly influenced by the *Volksinitiativen* as it is effectively impossible for any minority or collection of interest groups to directly impose their will on a majority of the populace. As a result, the entire political culture is strongly disposed towards compromise and consensus building. For example, the top executive decisions are made by a Federal Council of seven, who currently include members from four different parties, instead of having a single head of government.

The current strategy of the Swiss federal government is officially documented in the Energieperspektiven 2050 report (EP2050),[1] which essentially functions as a white paper (BfE, 2013). However, for all its length, the EP2050 report is vague on implementation: its main thrusts are efficiency, especially in buildings, e.g. replacing oil heating with more efficient heat pumps, further electrifying transportation, replacing Swiss nuclear power with Swiss domestic renewables, and a larger role for (existing) hydropower and rooftop solar photovoltaic (PV) panels. This new renewable electricity capacity is to generate 20% of Swiss supply by 2035, or around 12 TWh. The federal government is opting for natural gas power plants and imports of foreign renewables as a stopgap measure in case domestic renewable capacity does not expand fast enough.

The first phase of the Swiss ES2050 was passed into law by parliament in 2016. The Swiss parliament added several changes in the process, including a support package for hydropower, which has been economically less viable due to current low wholesale electricity prices. The new energy law was opposed and immediately challenged by one of the larger parties in parliament (see Überparteiliches Komitee gegen das Energiegesetz, 2016) but the Swiss voted to keep the law in a referendum (Der Bundesrat, 2017). The Bundesamt für Energie (BfE – Ministry for Energy) is currently working on the next phase of the ES2050.

If Switzerland is serious about replacing nuclear, it has four options available. The first is to use domestic renewables, mostly by expanding the number of solar panels and wind turbines. The second is to import renewables from foreign countries. These two seem the most likely and currently enjoy the support of a large majority of Swiss citizens in surveys (see following text). We will explore the risks and challenges that come with these options in detail over the next sections.

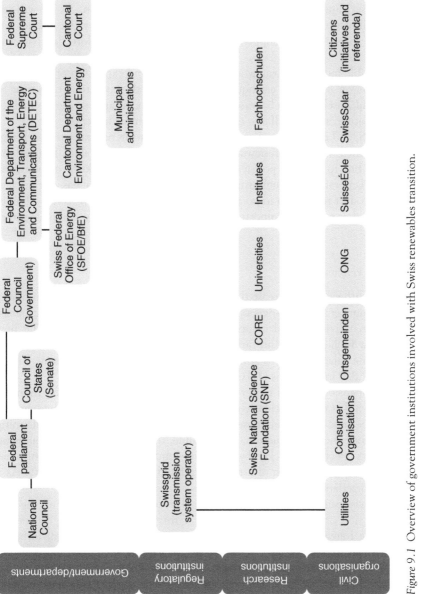

Figure 9.1 Overview of government institutions involved with Swiss renewables transition.

A possible third pathway is to use non-renewable natural gas. Gas is not currently used much for electricity in Switzerland (only for heating and chemical processes) but expanding its use is an option in the ES2050. The main risks for this pathway are that natural gas needs to be imported (e.g. from Russia) and the acceptability of increased air pollution. As we will see in the following sections, natural gas is neither preferred by the public, nor necessary for stability of supply. We therefore will not explore this pathway in depth.

A possible fourth pathway would be based on rapid expansion of (deep)geothermal power. However, earthquakes associated with geothermal tests near St Gallen and Basel seem to have soured the public mood against geothermal for the near future, and we will also not explore this pathway in depth.

The ES2050 emphasises domestic renewables, using natural gas as a bridging fuel until sufficient renewable capacity becomes available. Switzerland therefore has to make a political decision on whether to keep emphasising domestic generation (though using imported fuels like natural gas and nuclear fuel) or to start relying on imported electricity year-round. This is not a binary choice as domestic and imported renewables are potentially complementary pathways.

Domestic renewables

The first pathway is to increase the capacity for renewables production at home. A large number of PV panels could fit on existing rooftops in built-up areas, owned by individuals and companies that also own the buildings. The new energy law includes a feed-in tariff to support expansion of renewables but the Swiss home solar industry is still in its infancy compared to Germany, with a total of 49,000 installations (1.4 GW) completed by 2015.

The expansion of wind power in Switzerland has been even slower than PV because wind is not particularly strong and consistent in most of the country, and also because residents opposed to wind turbines have used spatial planning laws to block their construction. For example, the Kirchleerau–Kulmerau wind project was effectively blocked by a local referendum that ruled that no turbines may be built within 700 m of a residential building in that municipality. As a result, Switzerland had a total of 37 wind turbines in 2015.

The Swiss can also build some additional hydropower but nearly all the best potential sites are already in use. Much of this pathway therefore centres on small-scale renewables: rooftop PV, wind projects with a handful of turbines, and small hydropower plants. In turn, this means that much of the expansion will be in the hands of private citizens, small and medium-sized businesses, and municipalities. Federal and cantonal authorities can support this development, in part through legislation that favours renewables and policies of the utilities they (jointly) own.

The two main risks to the domestic renewables pathway are the intermittency of solar PV and wind power in Switzerland – which does not have a lot of sites with strong and sustained sunshine or winds – and public acceptance of the large number of PV panels and wind turbines necessary to make this work.

Foreign renewables

The second pathway is to import renewable electricity from abroad. This would require a build-up of renewable capacity outside Switzerland to feed into the Swiss grid. There are two general options: wind power, most likely from the west coast of Europe like the North Sea, or concentrating solar power (CSP), from sunny regions like southern Spain or North Africa. This would let the Swiss use the most abundant renewable resources in and near Europe without having to build in populated areas, making for cheaper electricity.

Imports are feasible from a grid perspective: Switzerland already has the transmission lines and interconnections to move large volumes of electricity. These are currently used for three things: exporting excess electricity in summer, importing electricity in winter, and transporting electricity from north-west Europe to Italy. Both imports and exports are equivalent to 75% of gross national electricity production. Despite this trade, Switzerland sees itself as self-sufficient, with a net electricity export just under 2% of its total production in 2015 (BfE, 2016). However, the problem for this pathway is that these power plants are very far from the cities and other demand centres that they are to supply. Building the infrastructure needed requires action from larger utilities and co-ordination between the European grid operators, i.e. the members of the European Network of Transmission System Operators for Electricity (ENTSO-E). Swiss utilities are already investing in foreign renewables projects but this would need to be scaled up. Simultaneously, grid operators are in charge of constructing transmission lines to connect this new supply. This has not always gone smoothly; a major high-voltage line that was to connect wind farms in the Baltic with cities and factories in southern Germany has been opposed by residents and NGOs because of the landscape impacts and alleged health impacts it would have. Furthermore, it runs close to the eastern border and some have alleged it was planned to also carry Czech nuclear power, which seriously undermined the green narrative that supported this transmission line.

The main risks to the foreign renewables pathway are the construction time and vulnerability of the transmission infrastructure – the latter more due to extreme weather than terrorism – and the acceptability to the Swiss public of foreign control over power supply. A further risk is in the acceptability of the power plants and transmissions lines for foreign benefit to the residents of other countries. Assuming we focus on the offshore wind and CSP plants in sparsely populated areas, which are anyway more productive, international transmissions lines face the largest risk of popular resistance.

Research process and methods

Four research methods inform this narrative: energy system modelling, Q-methodology, a choice experiment, and a stakeholder workshop.

The core of energy system modelling is to represent major supply-and-demand technologies in our energy system, and the flows between these, in one single model. The mix of these technologies is constrained by existing real-world circumstances and scenario assumptions, and can be optimised within these constraints, e.g. for lowest total cost given high reliability and low CO_2 emissions. We looked specifically at electricity and included various supply technologies, including run-of-river and dam hydropower, PV and CSP, wind turbines, gas turbine power plants, and pumped storage. Our model, using the Calliope framework (Pfenninger, 2017) represents sources in Switzerland and abroad. We use hourly data for intermittent sources like wind and solar from www.renewables.ninja, a website with open renewable energy potential data (Pfenninger and Staffell, 2016; Staffell and Pfenninger, 2016) to find how intermittency limits the renewable electricity sources we can reliably use together, and how much we would have to use other sources as a backup to balance supply and demand.

The core of Q-methodology is the development of a set of statements expressing potential stakeholders' attitudes and beliefs about a particular issue, ensuring coverage and balance of the topic (Watts and Stenner, 2012). We started by interviewing a diverse range of stakeholders and systematically reviewing mass-media coverage. A sample of participants then ranked the statements according to whether they agree or disagree with their own perspectives. Using factor analysis, we identified patterns in the ranking and identified groups of participants who are likely to rank the various statements differently. In this way, we identified correlations and/or major differences between group perspectives. We could then match these perspectives to any shared institutional and/or demographic attributes of the groups' members. Q-methodology is used to explore perspectives of a small group of individuals that should not be interpreted as representative of a larger populace without further research to confirm it. We used Q-methodology to unravel stakeholder perceptions in three Swiss villages.

The core of a choice experiment is to make explicit what people base their decision on (Alriksson and Öberg, 2008). During the experiment we give respondents a set of different choice tasks, so-called 'choice sets'. Within each choice set, respondents chose their preferred option from three alternatives, one of which was a status quo option. It is assumed that the individual utility of a choice depends (in part) on different observable attributes that characterise the options within the choice sets. In our experimental design, we estimated part-worth utilities of the choice attributes by decomposing respondents' answers as well as other variables. We could also group respondents based on the similarity of their preferences using principal component analysis.

The core use for a stakeholder workshop is to make explicit the viewpoints of the different participants and facilitate an exchange of opinions and arguments between them. This may in turn create a shared vision but a more important goal is to build mutual trust and understanding for different viewpoints. Our workshop used a role-playing format, where we asked stakeholders to reason from the point of view of the voter groupings we found in our survey, to emphasise understanding other viewpoints and stimulate out-of-the-box thinking. We invited representatives of different governments, NGOs, utilities, and consultancies.

Risks and uncertainties

Risks to the Swiss renewables transition emerge at two levels. The first level is the new risks that replacing nuclear with renewables brings: intermittency, when no electricity can be produced as the sun is not shining, the wind is not blowing or rivers are frozen; failures in our electricity grid from extreme weather events such as from storms, icing, and landslides; and public opposition to new energy installations such as solar panels, wind turbines, and power lines.

Any combination of intermittency and grid failure may lead to power losses and blackouts in a country that is used to an extremely reliable supply of electricity. The pathways that we examine trade off these two risks: wind turbines in Switzerland produce less than half of what they do in Denmark (see renewables. ninja) but are close to the Swiss power grid; solar power from North Africa has very predictable output but needs long and vulnerable transmission lines that cross several countries.

For solar farms we find inherent tension between using land for nature or agriculture and infrastructure. This applies especially in Switzerland as much of its mountainous terrain is ill suited to infrastructure and the rest of the country is fairly densely inhabited. Furthermore, the need for permits and grid connections makes solar farms unattractive to utilities at current bulk electricity prices. By contrast, rooftop PV and solar heaters only need to compete with residential electricity prices, which are higher due to inclusion of grid fees and taxes. Installing PV on non-building infrastructure, like avalanche protection barriers, is possible, but such projects are still experimental and expensive.

The second level of risk is in the aggregate, or how the risks for the individual projects and technologies affect the overall Swiss energy strategy. While individual renewable projects may supply intermittent power or get disconnected, combining many different sources and existing Swiss hydropower can lead to a stable supply. However, public support is more difficult at the project level than in aggregate: on the national level, this is just an abstract percentage of supply, while on the cantonal level it is a question of where the infrastructure will be located, and on the local level it is a binary choice of having the infrastructure in your back yard or not. This is particularly fraught because local residents may experience fewer of the benefits and more of the drawbacks of a project that benefits the country as a whole.

Consequential risks (negative impacts)

Switzerland has one of the most reliable electricity supplies in the world right now, and its inhabitants see this as the right and proper natural state of things. Existing proposals implicitly or explicitly commit to operational security of supply, suggesting that the Swiss are unlikely to compromise on reliability for the sake of independence, climate, or a nuclear phase out. This stability can also not come at unlimited cost. We see two risks that would cause renewable electricity to lead to an unstable supply of electricity: intermittency and grid failure.

On intermittency, we find that Switzerland can phase out nuclear and switch to renewables without risk of intermittency as long as we do not rely solely on PV, even though there is sufficient space to put the panels (for more details, see Díaz Redondo, van Vliet, and Patt, 2017). This is because Swiss hydropower can compensate for limited intermittency, but less so in winter when PV output is low and Swiss rivers get limited water. Relying only on PV will require some seasonal electricity storage, which is currently prohibitively expensive due to the staggering volume of electricity it would have to store for every winter. Wind power from the North Sea is especially suited to the Swiss electricity system as it is more stable than solar power and produces more electricity in winter.

Furthermore, both North Sea wind and North African CSP are likely to be cheaper than using natural gas due to the rapid decline in installation cost for renewables electricity. As a measure of this, we use the levelized cost of electricity (LCOE) that is defined as the cost of the entire electricity supply system, including grid and backup plants to guarantee constant supply, divided by the kWh of electricity it supplies (in Swiss francs per kilowatt hour). For example, based on cost projections for wind and CSP from 2011 to 2016, replacing nuclear with a combination of wind and CSP would cause an LCOE ranging from about the same as using natural gas to almost twice as much. However, using commercial costs for wind and CSP contracted in 2017 results in an LCOE below these ranges. This showed that: (a) their model calculations were outdated by the time they were published; and (b) that renewables have reached 'grid parity'. The cost of generation is no longer a reason to avoid a switch to renewables.

If Swiss utilities can invest in, buy a majority stake, or otherwise gain control over one or two dozen wind farms and/or CSP plants abroad, this system would insulate Swiss electricity supply from the intermittency of individual renewable power plants and the resulting fluctuations in prices on power markets. This would be a shift in policy for Swiss utilities: they already own stakes in power plants in foreign countries (overwhelmingly in EU member states) but the electricity is sold on local markets, not imported back to Switzerland.

However, the second risk is that long power lines come with increased chance of outages due to extreme weather, which currently accounts for almost half of all grid outages. The magnitude of this risk depends on the grid: for a future with a large share of imports, sufficient redundancy in transmission corridors, high-quality equipment, and best practices in grid management can minimise the risk. Quantitative analysis and interviews with grid experts have shown that grids can almost always be built to withstand the harshest conditions in any given country. For example, the Finnish grid suffers more outages in the comparatively mild summer than in the harsh Nordic winter. Exchanging best practices would help transmission system operators (TSOs) prepare for changes in weather conditions brought on by climate change. This can be organised through existing organisations like ENTSO-E or Eurelectric. Furthermore, even if the transmission grid breaks down due to weather or for other reasons, Switzerland has a large capacity for hydropower to provide some short-term buffer.

Our exploration of these two consequential risks shows that both can be mostly avoided and a fully renewable electricity supply for Switzerland can be realised at reasonable costs without an increased risk of supply interruptions. The only option that does not work is to rely on PV only. Without overwhelming technical or economic constraints, this means the choice between pathways is essentially a matter of public preference. And here we find that it is not immediately clear that any of these supply options can be realised.

Implementation risks (barriers)

The implementation barriers we find for the Swiss ES2050 are of a social and political nature rather than, for example, lack of access to financing. For example, some legal bottlenecks need to be sorted out, such as cost-sharing or compensation for people who rent and wish to install PV on the roof of the building they live in. This particular problem of agency is especially acute in Switzerland as 80% of the population live in rented houses and apartments, and it requires a legal solution.

However, the largest perceived barrier to renewables in Switzerland is public acceptance. The prevailing view is that the high attachment of the Swiss to their traditional landscape, and the large number of landscape and nature NGOs, make the construction of power lines, wind turbines, PV farms, and conspicuous rooftop PV in historic centres likely to attract opposition. Several research projects are underway to address this directly, and most of the projects in NFP70, a research programme about electricity supply options for Switzerland, include an 'acceptance' component in addition to the technical research at the heart of these projects.

We could obviate acceptance issues in Switzerland in theory by outsourcing electricity supply to other (neighbouring) countries and the Swiss transmission grid could most likely handle the import load. However, this leads to two potential problems.

First, citizens in these countries may be equally or even more unhappy to have energy infrastructure in their environment to supply someone else, though this might be offset by the business opportunity of selling the electricity. Moreover, if all of Europe follows in the footsteps of a renewable Switzerland, the required international interconnect capacity would need to grow by a factor of 6 to 12 (Rodríguez *et al.*, 2014). Expansion of transmission lines would especially be needed in areas where renewable electricity is generated, which are usually peripheral areas.

Second, this creates a reliance on other countries. The Swiss have a strong interest and desire for energy independence, often expressed as a desire for electric autarky (Trutnevyte, 2014). This has some dissonance with the fact that Switzerland currently imports 75% of its energy in the form of natural gas, oil, and uranium to fuel nuclear plants (BfE, 2016), and also with recent efforts by utilities to invest in renewable electricity generation abroad. Regardless of the current realities of energy use, independence is an aspiration that makes imports, renewable or fossil, less politically attractive.

Swiss citizens consistently prefer solar electricity and, to a lesser degree, wind (for more details, see Plum *et al.*, n.d.). These should be built in existing industrial and commercial areas, including ski resorts, rather than in areas of natural beauty. However, the Swiss do not agree on all aspects of their energy strategy. We find five distinct groups of voters: the largest two groups are 'Moderates' and a group that is specifically 'Contra status quo' to using nuclear power but otherwise also moderate. Three groups have a very specific profile: 'Pro renewables', 'Pro Switzerland', and 'Pro landscape'. All groups except 'Pro landscape' (95% of respondents) prefer electricity from Switzerland, and all groups except 'Pro Switzerland' (84% of respondents) accept imports of renewable electricity, preferably from plants operated by Swiss utilities. Unlike domestic or imported renewables, Swiss generally dislike natural gas and non-renewable imported electricity. Furthermore, our survey results show that the Swiss public find 'the construction of high-voltage transmission lines abroad for supplying Switzerland with electricity problematic' just as much as having power lines or wind turbines in their own living environment.

Local perspectives

Three local renewable energy projects in Switzerland illustrate the perspectives of different stakeholder groups on the process and negotiations involved in actually implementing a renewable energy project and show approaches to manage the risks involved: a small hydropower project in the canton of St Gallen (examined in Díaz Redondo, Adler, and Patt, 2017), solar PV on avalanche barriers in the canton of Graubünden (examined in Díaz Redondo and Van Vliet, 2018) and a solar farm project in the canton of Vaud (examined in Späth, 2018).

The small hydropower project in Breschnerbach (canton St Gallen) was led by a local utility, consultants, and pro-hydropower members of local and cantonal governments. The main challenge noted by these proponents was that the decision-making process, including permitting, took a very long time to complete. In the event, NGOs in favour of reducing energy demand were in conflict with local proponents who saw the hydropower plant as benefiting the local economy and community. After the utility received the concession in 2011, four years were spent negotiating a nature compensation scheme, with some 40 stakeholders represented in the process, including national NGOs. One NGO complained to the cantonal court but this was rejected, and the construction permit was granted in 2016. The project was eventually seen as 'win–win'. All perspectives showed consensus on the need for fair, inclusive, and democratic decision making, though the local proponents felt that participatory decision making limits infrastructure development.

In St Antoniën (canton Graubünden), a project aimed to install solar PV on avalanche protection barriers. This would generate more electricity than PV in lowlands as the solar input is up to 50% higher in the mountains. Moreover, suitable foundations were already present, and foundations make up a significant share of the installation costs for most PV projects. However, building PV on avalanche barriers in a remote village had never been done on a commercial scale. Everyone involved thought the project could contribute to the regional and local economy and benefit the environment. We found this to be the major driver for stakeholders to engage in the implementation process. Furthermore, the project was strongly driven by the idea that this could be a model for PV systems on avalanche defences. Most of the funding required for the project was crowd-sourced, with some contribution from the cantonal government. However, the federal government refused to fund the project as an innovation pilot and the municipal assembly ultimately voted against a loan to cover the funding shortfall. Residents, NGOs, and government officials suggested this was due to doubt over whether sales of electricity could cover investment costs, as well as their worries about estimates of technical difficulties, rising costs during the decision process, and unsuccessful fundraising activities. This seems to have increased doubts about and opposition to the project, and eventually led to an opposing vote in the municipal assembly. Some of the stakeholders who were opposed also felt that they did not receive the information they wanted from the proponents of the project, and that this created a lack of trust. Better communication might have removed this risk. Furthermore, the decision of the federal government not to fund this rural PV project was seen as inconsistent with the national policy to promote rural development and renewable energy.

In Payerne (canton Vaud), the local utility proposed a solar farm on a plot of agricultural land that had already been designated as an industrial area. This project was part of a greater plan to make the village largely self-sufficient in electricity. Local stakeholders and NGOs were involved in the planning process and, while one NGO formally opposed the project, it did not take the project to the courts. The project was completed in 2015 and produces around 40% of the electricity used in the village. Stakeholders generally agreed that large roof surfaces should be used first to install solar panels, but there was a gap between traditionalists who wanted to reserve farmland for agriculture and pragmatists who will use it for solar PV if that is more profitable. Others emphasised energy efficiency, much like in Breschnerbach, and citizens' role in decision making, much like in St Antoniën.

Broader implications

The potential for utility scale PV is likely quite limited in Switzerland, given the value put on rural landscapes and agriculture, but fortunately the available rooftop area for PV seems sufficient for Swiss purposes in several estimates (see Gutschner *et al.*, 2002; Compagnon, 2004). Furthermore, permitting for any energy infrastructure is known to be a long process both inside and outside Switzerland. One utility noted that they only invest in foreign renewables projects that have already obtained permits in order to reduce their exposure to acceptance risk.

Utilities and renewable plant developers are particularly concerned about the current low price of electricity in Switzerland and the EU power market, as a result of the ongoing expansion of renewables in Germany and elsewhere. The 'merit order effect' reduces prices when renewable electricity is abundant (i.e. sunny weather and high winds across the EU) and drives up prices of balancing power when renewable electricity is scarce, making renewable electricity a victim of its own success (see Cludius *et al.*, 2014). This has reduced enthusiasm for installing renewable electricity by utilities. Consumer-owned rooftop PV has not been affected as much, as domestic production reduces electricity bills that include taxes and grid fees (electricity cost is only half the total) and the Swiss feed-in tariff policy.

Across the three projects, stakeholder perspectives suggested a risk to the overall Swiss ES2050: stakeholders feel disengaged with the ES2050 process because they disagree on the overarching framework, i.e. the pillars of the energy strategy. It seems that the federal, cantonal, and local stakeholders have different interpretations of the ES2050 and its major objectives of energy efficiency, supply diversity, deployment of investments, and environmental protection. Each of these levels of government prioritises the ES2050 objectives differently, and all of them seem to think their approach is best for everyone. However, these authorities lack a forum to resolve these differences. Furthermore, these interpretations differ in turn from the preferences of the Swiss public. This is a risk to the general political process that carries the ES2050 forward.

This is separate from the 'usual' political risk where different interest groups want different things. This is also present in Switzerland, and some of the lobby groups have very close ties to political parties.

Comparing pathways

Unsurprisingly, we cannot have an energy system that is reliable, climate friendly, gives us independence, keeps the landscape intact, and phases out nuclear all at the same time. However, it is possible to replace the existing nuclear plants in Switzerland with a combination of domestic and imported renewables without infringing on reliability. The cost for this would be no higher than using natural gas, as suggested in the ES2050, or replacing the ageing Swiss nuclear plants. Good management by the TSOs that carry the imported power would minimise the risk of weather-induced grid failures. Both pathways, domestic renewables and the foreign renewables, are therefore possible in principle, though rooftop PV would have to be supplemented with Swiss wind power in strictly domestic pathways. Due to the low potential for wind power in Switzerland (i.e. wind blows slowly, infrequently, and/or erratically), this would also have highest cost. Natural gas is technically feasible but a non-starter in the opinion of the Swiss public.

While both pathways are technically feasible and broadly socially acceptable, there are still issues with individual projects (see Table 9.1). As long as renewables partially depend on subsidies to be competitive, financing remains difficult and the projects will be seen as risky by investors. This is especially the case if

Table 9.1 Risks to successful expansion of renewable sources of electricity in Switzerland (CH)

Barriers	Hydropower in CH	Rooftop PV in CH	Utility scale wind/ PV in CH	Offshore wind	Mediterranean CSP
High investment costs	Compounded by low prices	Ameliorated by FIT and fees	Compounded by low prices	Compounded by low prices	Compounded by low prices
Landscape/visual impact	Only for large hydro	Minor	Unwanted	Low population if far offshore	Low population in desert
Permitting	Nature compensation done	Building owner's permission needed	Challenges expected	Transmission line may be challenged	Transmission line may be challenged
Intermittency	Base load and dispatchable	Diurnal, weather, and seasonal	Diurnal, weather, and seasonal	Relatively stable, better in winter	Thermal storage reduces
Energy independence	Swiss source	Swiss source, need balancing	Swiss source, need balancing	Contributes to diversification	Contributes to diversification

nature compensation adds to the cost. Delays in permitting and constructing all of the necessary power plants and transmission lines are also very likely. This is not necessarily a problem to either pathway in the long run, as the Swiss nuclear plants are scheduled to remain in service for as long as they are deemed safe to operate. As the two youngest and largest plants (built in 1979 and 1984, with a combined capacity of 2.2 GW) have recently received extensive safety upgrades in the wake of Fukushima, we can expect their lifespan to be extended to 60 years (World Nuclear Association, 2018). If these nuclear plants remain in service until 2039 and 2044, the Swiss have some 25 years to construct their replacements.

Some uncertainties remain in the Swiss transition to renewable electricity, although they are not threatening: For example, the political winds may shift over the next 25 years to be more or less open to international co-operation. However, the structure of the Swiss Confederation's executive makes dramatic shifts in policy – like in countries with two-party systems (e.g. the USA) – unlikely. Overall, we may expect Switzerland to remain rhetorically independent and autarkic, but practically integrated in Europe. Likewise, the Swiss economy has been relatively stable, with a largest annual drop of 3.4% of GDP in the last financial crisis, despite being known for international banking. The effects of climate change are also uncertain but the effects on hydropower are expected to be small over most of this century (SGH and CHy, 2011). Finally, it is unclear how far the costs of PV and wind will decrease in the future, but grid parity has now effectively been achieved and any further drops can only be in favour of the wider adoption of renewable electricity sources.

Conclusions

Overall, the Swiss prefer domestic production of renewable electricity, but a majority share of imported renewable electricity will likely be cheaper overall and cause fewer issues with intermittency. However, the renewable imports pathway would face more problems with acceptance of new infrastructure, especially long-distance transmission lines. (See Figures 9.2 and 9.3.)

The most recommended option would be to combine the domestic renewable pathway with the imported renewable pathway. The most favourable combination seems to be Swiss rooftop PV, offshore wind from the North Sea, and Swiss hydropower. Such a mix would also be acceptable to the Swiss public. This is especially important given the Swiss political system in which policies and projects can be challenged in local, cantonal, or national referenda.

However, depending on the demand for renewables in EU countries, this may require expansion of transmission capacity in the Dutch, Belgian, Danish, and German grids. Both the needs for grid expansion, and ways that this could be done in a manner acceptable to residents around the new transmission lines, should be researched further.

This narrative has two major implications for the Swiss energy strategy. First, the ES2050 can be broadened to include imports of wind and/or CSP, but Swiss

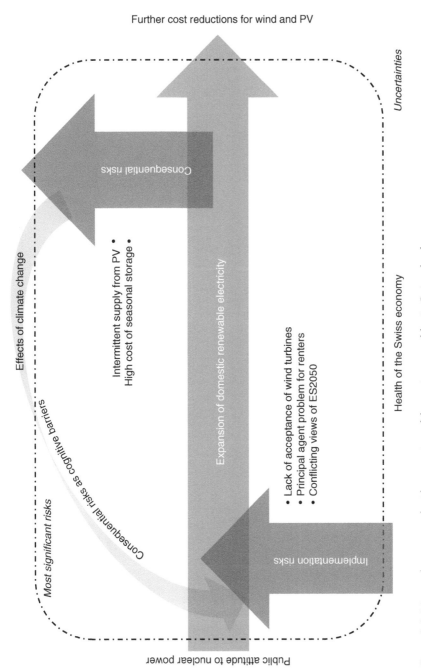

Further cost reductions for wind and PV

Effects of climate change

Uncertainties

Most significant risks

Consequential risks as cognitive barriers

Consequential risks

Intermittent supply from PV •
High cost of seasonal storage •

Expansion of domestic renewable electricity

Health of the Swiss economy

Implementation risks

• Lack of acceptance of wind turbines
• Principal agent problem for renters
• Conflicting views of ES2050

Public attitude to nuclear power

Figure 9.2 Main risks associated with expansion of domestic renewables in Switzerland.

Note

Assumes no significant changes in public attitudes to nuclear power, effects of climate change, downturn in the Swiss economy, and ongoing gradual reductions in costs of wind and PV installations.

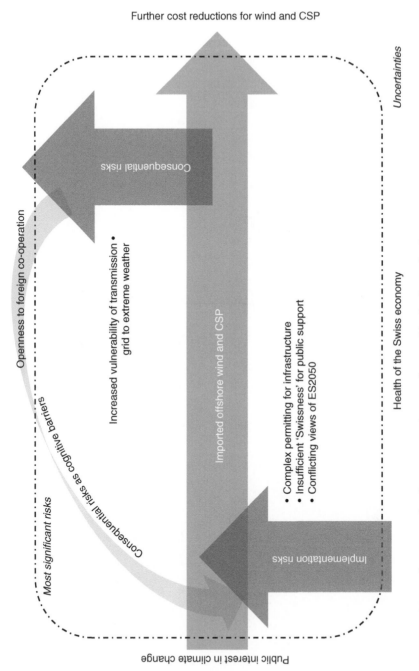

Figure 9.3 Main risks associated with expansion of imported renewables for Switzerland.

Note

Assumes no significant changes in public interest in climate change, openness to imported electricity, downturn in the Swiss economy, and ongoing slow reductions in costs of wind and CSP plants.

ownership and operational control would be preferable to the Swiss people. Second, as long as there is no forum to resolve the diverging interpretations of ES2050 among local, cantonal, and national stakeholders, we can expect conflicts and delays.

The risks we examined were of a political, technical, and economic nature. The political risks were mostly barriers to implementing one of the pathways, and the technical and economic risks were mostly about the consequences of these pathways. This follows a pattern we have observed in general in the literature about the Swiss energy transition.

The most pressing risk seems to be delay or outright failure to obtain permits, and more generally how to plan and build energy infrastructure without provoking opposition and legal challenges from nearby residents. This has been done successfully in Switzerland, for example for the Linth–Limmern pumped storage plant and its connection to the grid, where residents raised no objections. It would be worthwhile to investigate successful processes for energy infrastructure and determine how these can be mainstreamed.

Note

1 The EP2050 was written by Prognos AG, the consultancy firm that also wrote Energiekonzept 2050, a similar document, for the German government.

References

Alriksson, S., Öberg, T. (2008). Conjoint analysis for environmental evaluation. *Environmental Science and Pollution Research* 15(3), 244–257. doi: 10.1065/espr2008.02.479.

BfE (2013). *Energieperspektiven 2050*. Bern. Available at: www.bfe.admin.ch/themen/00526/00527/06431/index.html?lang=de.

BfE (2016). *Gesamtenergiestatistik 2015 Tabellen*. Bern. Available at: www.bfe.admin.ch/dokumentation/publikationen/index.html?start=0&lang=en&marker_suche=1&ps_text=Gesamtenergiestatistik&ps_nr=&ps_date_day=Tag&ps_date_month=Monat&ps_date_year=2012&ps_autor=&ps_date2_day=Tag&ps_date2_month=Monat&ps_date2_year=Jahr&ps.

Der Bundesrat (2017). *Energiegesetz (EnG)*. Available at: www.admin.ch/gov/de/start/dokumentation/abstimmungen/20170521/Energiegesetz.html.

Cludius, J., Hermann, H., Matthes, F.C., Graichen, V. (2014). The merit order effect of wind and photovoltaic electricity generation in Germany 2008–2016: Estimation and distributional implications. *Energy Economics* 44, 302–313. doi: 10.1016/j.eneco.2014.04.020.

Compagnon, R. (2004). Solar and daylight availability in the urban fabric. *Energy and Buildings* 36(4), 321–328. doi: 10.1016/j.enbuild.2004.01.009.

Díaz Redondo, P., van Vliet, O.P.R. (2018). Drivers and risks for renewable energy developments in mountain regions. A case of a pilot photovoltaic project in the Swiss Alps. *Energy, Sustainability and Society* 8(28). doi: 10.1186/s13705-018-0168-x.

Díaz Redondo, P., Adler, C., Patt, A. (2017). Do stakeholders' perspectives on renewable energy infrastructure pose a risk to energy policy implementation? A case of a hydropower plant in Switzerland. *Energy Policy* 108, 21–28. doi: 10.1016/j.enpol.2017.05.033.

Díaz Redondo, P., van Vliet, O.P.R., Patt, A.G. (2017). Do We Need Gas as a Bridging Fuel? A Case Study of the Electricity System of Switzerland. *Energies* 10(7), 861. doi: 10.3390/EN10070861.

Gutschner, M., Nowak, S., Ruoss, D., Toggweiler, P., Schoen, T. (2002). *Potential for Building Integrated Photovoltaic.* T7–4. St Ursen, Switzerland. Available at: www.iea-pvps.org/index.php?id=9&eID=dam_frontend_push&docID=394.

Pfenninger, S. (2017). Dealing with multiple decades of hourly wind and PV time series in energy models: A comparison of methods to reduce time resolution and the planning implications of inter-annual variability. *Applied Energy* 197, 1–13. doi: 10.1016/j.apenergy.2017.03.051.

Pfenninger, S., Staffell, I. (2016). Long-term patterns of European PV output using 30 years of validated hourly reanalysis and satellite data. *Energy* 114, 1251–1265. doi: 10.1016/j.energy.2016.08.060.

Plum, C., Olschewski, R., Jobin, M., van Vliet, O.P.R. (n.d.). Public preferences for the Swiss electricity system after the nuclear phase-out: A choice experiment. *In review.*

Rodríguez, R.A., Becker, S., Andresen, G.B., Heide, D., Greiner, M. (2014). Transmission needs across a fully renewable European power system. *Renewable Energy* 63, 467–476. doi: 10.1016/j.renene.2013.10.005.

SGH and CHy (2011). *Auswirkungen der Klimaänderung auf die Wasserkraftnutzung – Synthesebericht* 38. Bern. Available at: https://naturwissenschaften.ch/service/publications/52708-auswirkungen-der-klimaaenderung-auf-die-wasserkraftnutzung-synthesebericht.

Späth, L. (2018). Large-scale photovoltaics? Yes please, but not like this! Insights on different perspectives underlying the trade-off between land use and renewable electricity development. *Energy Policy* 122, 429–437. doi: 10.1016/j.enpol.2018.07.029.

Staffell, I., Pfenninger, S. (2016). Using bias-corrected reanalysis to simulate current and future wind power output. *Energy* 114, 1224–1239. doi: 10.1016/j.energy.2016.08.068.

Trutnevyte, E. (2014). The allure of energy visions: Are some visions better than others? *Energy Strategy Reviews* 2(3–4), 211–219. doi: 10.1016/j.esr.2013.10.001.

Überparteiliches Komitee gegen das Energiegesetz (2016). *Referendum gegen ruinöses Energiegesetz.* Available at: https://energiegesetz-nein.ch [accessed 10 November 2016].

Watts, S., Stenner, P. (2012). *Doing Q Methodological Research – Theory, Method & Interpretation.* SAGE Publications, London, UK. Available at: https://uk.sagepub.com/en-gb/eur/doing-q-methodological-research/book234368.

World Nuclear Association (2018). *Nuclear Power in Switzerland, Country Profiles.* Available at: www.world-nuclear.org/information-library/country-profiles/countries-o-s/switzerland.aspx [accessed 9 August 2018].

Part IV

Pathways towards energy-efficient building sectors

10 China

Risks and uncertainties in low-carbon pathways for the urban building sector

Lei Song, Jenny Lieu, and Ying Chen

Introduction

China has become both the world's largest energy producer and consumer. It has a comprehensive energy supply system based on coal, electricity, oil, natural gas, and renewable energy, meeting the basic demands for socio-economic development. However, China's energy production and consumption are also facing difficult decarbonisation challenges due to rapid economic development, abundant endowment of coal, population growth, and urban development.

The country has experienced the fastest economic expansion of any major economy in the world and has taken 500 million of its population out of poverty. Today, China is the second largest economy in the world, with an overall GDP of ¥67.67 trillion (US$11.06 trillion) in 2015 (NBS, 2016a). Economic growth is now slowing, falling to 6.9% in 2015 compared with the average rate of 9.7% from 1979 to 2014. During the 13th Five-Year Plan (2016–2020) period, China's potential average growth rate is expected to drop to 6.3%, indicating that China's economy is transitioning from high to medium-high growth.

This growth has been accompanied by increasing energy demand. Even though the energy supply and consumption structure has been improving in China, the long-term socio-economic development will continue to rely on coal, of which China has large resources, resulting in increasing carbon emissions. Energy consumption in building, transport, and daily living sectors in particular will keep on increasing. As of 2014, transport and housing have become the largest growth areas for energy consumption in China (Wei and He, 2017).

China's population exceeded 1.38 billion in 2016. Urbanisation levels reached over 57% – around 3% higher than the global average (NBS, 2016a). Although the speed of urbanisation has slowed somewhat – down from nearly 4% in 2010 to around 2.5% in 2016 – the urban population is expected to reach 65% by 2020. Along with urban population growth, rising living standards in cities, locked-in energy infrastructure, and transport-related activities will contribute significantly to increasing carbon emissions. Cities will therefore need to make major carbon emission reductions in the present and future.

Increasing urban population and density have driven both the rapid development of the building industry and rising energy consumption in China. In 2015, the construction industry contributed 26.4% of China's GDP, and building energy use accounted for 33% of the total energy use in China (Yuan *et al.*, 2017). Total carbon emissions from China's building sector increased from 984.69 million tons of CO_2 in 2005 to 3,753.98 million tons of CO_2 in 2014 (Jiang and Li, 2017). Between 2001 and 2014, both primary energy consumption and electricity consumption in China's building sector increased more than two-fold (Li *et al.*, 2017) Rapid urbanisation and economic development have driven the demand for higher quality living spaces, including improved indoor comfort, with a corresponding increase in energy consumption. Therefore, energy conservation in new buildings and increasing demand in the existing building stock have become two of the largest challenges for China's energy conservation and emissions-reduction work (IEA and Tsinghua University, 2015). Building energy-conservation plays an important role in ensuring that emissions peak before 2030, a commitment set by the Chinese government in China's Nationally Determined Contributions (NDCs). Guidance and support from government policies are crucial for achieving the energy conservation targets for the building sector (Yuan *et al.*, 2017).

The low-carbon transition pathway for the Chinese building sector is determined by national regulations and plans related to building energy conservation. This narrative addresses two key components of a green building low-carbon transition pathway: energy-efficiency efforts to reduce emissions, and green energy consumption in buildings with an emphasis on scaling up renewable energy. ('Green buildings' are commonly understood as the practice of creating resource-efficient and healthier approaches for building design, construction, renovation, operation, and maintenance (Fastenrath and Braun, 2018).) Both pathways focus on urban residential buildings Although energy consumption in public and commercial buildings represents a dominant share in China's building sector – almost three times greater than residential buildings in rural or urban areas – rapid urbanisation has boosted continuous development and scaled up the construction industry, especially with regards to residential buildings. For new buildings in China, the construction areas of residential housing accounted for about 75% and for public buildings about 25% (NBS, 2016b). Furthermore, residential buildings in urban areas are now the highest energy consumers. For instance, space and water heating in urban buildings in northern China is the largest energy consumer, representing 52% of total building energy consumption in the region. In addition, demand for space heating and cooling for southern households in urban areas has been increasing rapidly due to climate change and higher living space comfort (Tsinghua University, 2016).

A low-carbon transition in residential buildings is more complicated than other building categories. This increased complexity is due to the wide range of housing types; differences in household demographics; varying climate conditions across China that impact households' ability to regulate indoor temperature; and economic-social contexts such as the housing rental market, social customs, and culture. This chapter illustrates how these factors may shape a low-carbon transition pathway in the building sector, the barriers to implementing the transition, and also the potential negative impacts resulting from the transition pathway.

Policies for a low-carbon transition in the building sector

China's low-carbon transition in the building sector is defined by an enabling set of overarching plans, objectives, and programmes. China's national medium and long-term technology development programme (2006–2020) proposes building energy-efficiency and green building as priority goals for a low-carbon transition in the building sector. Overarching commitments to the Paris Agreement are highlighted in China's NDCs, which also placed emphasis on enhancing green building in the whole life cycle of the building sector. Most recently, the National New-type Urbanization Plan (2014–2020) offered a lifestyle guideline that promotes 'green production and consumption to be part of the mainstream urban economic life'. The plan proposed to 'greatly increase the proportion of energy-saving and water-saving products, recycled products and green buildings' to achieve improved building energy conservation. China also issued a series of work plans that were closely related to building construction work, which sets a clearer vision for the pathway for low-carbon transition in the building sector from the supply side.

For promoting energy efficiency in the building sector, both new green building and retrofitting programmes were established. The 13th Five-Year Plan for Building Energy Efficiency and Green Building Development, released by the China State Council and the Ministry of Housing and Urban–Rural Development (MOHURD) in 2017, set up exact targets for increasing the proportion of green building and energy-efficient buildings in existing buildings stocks. By 2020, the energy efficiency level of new buildings in urban areas is planned to increase by 20% compared with 2015. The proportion of urban green building in new building developments is intended to be increased to 50%, while the proportion of green building materials used should exceed 40%. Over 500 million square metres of existing residential buildings should receive energy-saving renovations. Furthermore, the Energy Development Strategy Action Plan (2014–2020), released by the China State Council in 2014, has strengthened the actions required to achieve the target for controlling energy consumption and improving energy efficiency. This includes upgrading energy-efficiency design standards for residential buildings, accelerating the construction of green buildings and the renovation of existing buildings, setting up energy consumption limits for public buildings, a green building rating and labelling system, promoting energy appliances and green lighting, and reforming heat metering and transformation of existing buildings.

Several policies have been implemented to increase the use of green energy in the building sector. The 13th Five-Year Plan for Energy Conservation and Emission Reduction Programme, released in 2017, sets targets specifically for ultra-low energy consumption and zero energy consumption, including the use of distributed roof photovoltaic (PV) power generation systems, solar thermal, geothermal energy, and industrial waste heat to meet the demand of energy consumption in building sector. By 2020, the installation area of solar water heater systems will reach 45 million square metres, and the solar thermal utilisation for heating will reach 800 million square metres, which has been proposed in the 13th Five-year Plan for Renewable Energy Development issued in 2016. Additionally, there are regional plans that address energy demand in building. For instance, heating in the northern region of China represents 42.5% of Chinese final energy use in buildings during 2015 (IEA and Tsinghua University, 2015). Thus, the Clean Winter Heating Plan for Northern China (2017–2021), issued in 2017, required the northern regions (including Beijing, Tianjin, Hebei, Henan, and Shandong) to achieve a 'coal to gas' transition to provide half their power for heating from geothermal, biomass, solar, natural gas, electric, industrial waste, and centralised clean coal-fired heating by 2019. By 2021, the clean energy heating rate is expected to reach 70%.

In addition to national policies, in China local governments play substantial roles in promoting low-carbon development. For instance, the Green Building Evaluation Standard (GB/T50378–2014) was enacted in 2014 by MOHURD (the Ministry of Housing and Urban–Rural Development) to evaluate the conservation of energy, water, and materials in whole building life cycles. Local governments also issued their own standards for green building evaluation, such as Shanghai Standard for Green Building Evaluation (DG/TJ08–2090–2012) and the Beijing standard (DB11/513–2015). The local building codes may be based on the local development requirement, but these would be stricter than nationwide codes. Green building strategies from national and local level may be divided into five different categories such as: green building codes, retrofitting programmes, rating and labelling in green building, subsidy and preferential policies, and information campaigns to raise awareness.

Nested in the above strategies, the green building transition pathway consists broadly of two components: energy efficiency and renewable energy. The technologies, policies, and actions needed to promote energy efficiency and renewable energy are equally important for the transition towards a green building sector. However, here there will be a stronger emphasis on energy efficiency in the pathway description.

The green building pathway

Green buildings broadly focus on resource savings – with respect to energy, water, land, and materials – in the whole life of the building. The adoption of 'green' architectural principles such as solar, passive or low-energy design, and 'low-carbon' building technologies and materials generally reduces energy consumption and greenhouse gas (GHG) emissions (Pachauri et al., 2007; UNEP, 2011). Green building strategies may include different categories such as: certified green building labels, prefabricated buildings, new buildings with high energy efficiency, retrofitting existing buildings to high energy efficiency standards, renovating existing buildings with exterior windows or window shading, on-site renewable energy, green or vegetated roofs and walls, and building energy-saving management and services projects.

Given the context of China's low-carbon transition to green building in urban areas, the green building pathway consists of two main components: energy efficiency in the building sector, and renewable energy in the building sector and the electricity system. These have also been indicated in The Comprehensive Work Plan for Energy Conservation and Emission Reduction in the 13th Five-Year Plan Period. The component on energy efficiency of green building again contains two parts: new buildings and retrofitting existing buildings. Renewable energy in the building sector is not only about on-site renewable energy installations but also emphasises the promotion of renewable energy supply in the electricity system.

The green building transition pathway can be viewed through the life cycle of the building including: building-process energy consumption, which consists of building planning, design, and construction, and a building's energy consumption, consisting of the operation and utilisation of the building (Chen and Luo, 2008). Energy consumption during the operational phase accounted for 80% of total consumption during the life time of the building (Zhang, Kang, and Jin, 2018). Thus, a low-carbon transition that emphasises the operation phase of a building will be the primary emphasis in the pathway's discussion.

Research process and methods

The present chapter is based on a qualitative research design that includes two methodical approaches: stakeholder analysis, involving mainly workshops and interviews, and policy document analysis.

The data collection started with governmental documents, archives, and related literature reviews. This covered more than 50 documents issued by national governments and more than 19 regulations or programmes at the local level. These documents helped identify key factors, processes, actors, and the political framework of the building context. Based on that, a key set of questions was discussed with stakeholders in workshops and interviews, which developed a general understanding of transition pathway in building sector:

1 Is the current policy mix deemed adequate for achieving the desired green transition in the building sector? What additional policies are required to promote the transition pathway?
2 What are the barriers and challenges in the implementation of transitioning the building sector? What are the criteria from the stakeholders for estimating them?
3 What are the (positive or negative) impacts of green transitions on the social-technological structures?

Furthermore, the participatory research approach contributed to provide a multiplicity of opinions and perceptions about the drivers, implementation risks, consequential risks, and uncertainty under green building pathways in different climate zones.

Three workshops and two focus group interviews were held in Beijing and Shanghai respectively. The participants in each focus group were invited from different stakeholder groups including local governments, architects, constructors, NGOs, experts on green building design, urban planning and low-carbon technologies, and researchers. The events counted a total of 189 participants, including 132 governmental officials, 18 experts, six architects, six NGOs, four constructors, eight residents, and 21 researchers. The workshops helped to outline and specify the transition pathway, achieving a general understanding of the drivers for uptake by stakeholders and the barriers to a low-carbon transition. The interviews aimed to gain in-depth expert and practitioner knowledge. Most of the participants came from the local administration and experts in Beijing and Shanghai's building sectors, which identified the risk categories (implementation and consequential risks, as well as uncertainty) and their potential impacts under the transition pathway for the building sector.

Stakeholders for this research project were chosen as key informants in order to identify risks associated with low-carbon transitions in the building sector. Therefore, stakeholders were included who are involved in the building industries or engaged in the fields of policymaking, research, and consulting with respect to low-carbon development. These stakeholders are familiar with the low-carbon transition process in China. Policymakers play an important role in bringing about low-carbon transitions in China and consequently formed the largest stakeholder group. The opinions from the other stakeholder groups complemented the policymakers' perspectives.

Energy efficiency in the building sector

Energy efficiency in new green buildings

The transition pathway for China's building sector aligns with the goals set out in the 13th Five-Year Building Plan. By 2020, China envisions achieving a 20% improvement in energy efficiency for new buildings in urban areas compared to 2015. In less than two years, the proportion of urban new buildings defined as 'green buildings' shall be improved to 50% and the proportion of green building materials would exceed 40%. The green buildings targets shall mainly be achieved by upgrading the energy efficiency standards and certification system, which has been emphasised in the policies of 13th Five-Year Plan for Building Energy Efficiency and Green Building Development and the Energy Development Strategy Action Plan (2014–2020).

In the near future, new buildings will be constructed in line with energy efficiency standards such as the Three Star System certification standard. This certification system is expected to encourage technological innovation and improve the management of green buildings by means of two different certifications: design label and operation label. Design label is a pre-certification that can be granted to allow the project to market itself as a green building. Operation label is final certification that is granted after controlling the energy performance of the building for one year (Geng *et al.*, 2012). The Three Star System involves six indicators such as land saving and the outdoor environment, energy saving and energy utilisation, water saving and water resource utilisation, material saving and material resource utilisation, the indoor environment quality, and operating management. These indicators are intended to provide a strong signal for construction companies/developers to consider the energy and resource saving potential of their current designs. Eventually all construction companies/developers will reach minimum energy and resource efficiency requirements and may even exceed the standards to improve their competitive edge in a market that is becoming increasingly conscious of environmental impact and energy savings.

Furthermore, there is significant potential for the development of prefabricated buildings and promotion of green building materials as innovative solutions to meet the green building goal, which would increase 'passive' ultra-low-energy-use buildings. These measures have been strengthened in several policies such as the Action Plan for Urban Adaptation to Climate Change (issued by MOHURD in 2013), the Technical Guidelines for Passive Ultra-low Energy Use in Green Buildings (issued by MOHURD in 2015) and the 13th Five-Year Plan for Energy Conservation and Emission Reduction Programmer (issued in 2017). Since the majority of prefabricated buildings are largely manufactured in controlled settings, the process reduces environmental impact during the construction phase and results in savings (compared to traditional construction methods) in water consumption of around 50%, mortar consumption of 60%, wooden materials of 80%, and energy consumption of 20%. Although the cost of a prefabricated building is around ¥1000 per square metre higher than the cost of a normal building, they are still expected to account for 15% of new buildings until 2020 (MOHURD, 2017) as set out in the 13th Five-Year Building Plan's targets. Costs of prefabricated buildings in the short to medium term (2–5 years) are expected to decrease, through economies of scale and technological learning. In the future, the share of prefabricated buildings is expected to continue rising due to other advantages, such as reduced construction noise and dust in neighbourhoods and better-quality control of construction materials. These challenges are specifically linked to new builds, while the existing building stock faces a different set of challenges and should consider the needs of current occupants.

Retrofits of existing buildings

The low-carbon pathway for existing buildings is also based on the Energy Conservation and Emission Reduction Programme and the 13th Five-Year Plan for Building Energy Efficiency and Green Building Development. These programmes foresee that China will by 2020 complete the energy-efficient renovation of over 500 million square metres of existing residential buildings and the energy-saving reconstruction of 100 million square metres of public buildings.

Retrofits of existing buildings in China are expected to increase by 5% per year by 2020 and could continue to increase in the future. If followed through, the proportion of energy-efficient buildings in the existing urban residential building stock will reach 60% by 2020. Economically developed areas such as Beijing, Shanghai, Tianjing, Jiangsu, Zhejiang, Shandong, and Shenzhen have achieved breakthroughs in energy efficiency, and the proportion of energy-saving measures has exceeded 10% of local green building polices across China (Chen, 2018).

In practice, renovation projects for existing buildings involve three components: the external structure, the internal equipment, and the renewable energy supply. With the exception of renewable energy use in existing buildings (which will be discussed in the next section on green energy), the external structure retrofits include external wall and roof insulation for improving thermal resistance or heat storage capacity, and energy-efficient doors and windows with low-emissivity glass. For instance, the Action Plan for Urban Adaptation to Climate Change (released in 2013) proposed to accelerate the comprehensive transformation of aged residential buildings by improving the buildings' air tightness and strengthening their performance in water collecting and heat insulation. Furthermore, the national plan also mentions raising the renovation standard of energy and water saving for existing buildings. Internal equipment retrofits can promote energy-saving products for HVAC (heating, ventilation, and air conditioning), lighting, and operation controls such as Building Information Modelling (BIM) or energy monitoring systems (Li *et al.*, 2017). Some related measures to renovate the existing buildings have been mentioned in the Energy Development Strategy Action Plan (2014–2020), such as promoting energy-efficient appliances and green lighting in existing buildings, and accelerating the transformation of existing buildings by the reform of heat metering and charging. Furthermore, energy efficiency improvements are expected to continue to increase as retrofits improve energy efficiency along with associated behavioural changes. Households will become increasingly aware of the economic and environmental benefits of energy efficiency. There may also be positive reinforcement of energy-efficient behaviours based on the piloted Social Credit System, which can rate consumer behaviours based on data collected.

There will also be continued technical support and financial incentives (including subsidies and green financial bonds) for retrofits of existing buildings that comply with increasingly stringent energy codes, especially in residential buildings in northern China and in public buildings nationwide (Yu, Evans, and Shi, 2015). The Regulation on Reward for Heat Supply Metering and Energy Conservation of Existing Residential Buildings in North Heating Areas (released in 2007) and the Regulation on Subsidy Funds for Energy-saving Renovation of Existing Residential Buildings in Hot Summer and Cold Winter Zone (in 2012) have stated that central finance allocate special funds or rewards to support energy saving in existing residential buildings in different climate zones, such as the northern and southern regions. More specifically, subsidies in Shanghai that were first provided in 2013 will continue to support retrofits, renovations, and/or monitoring energy use in existing buildings over the next decade.

Green energy in the building sector and the electricity system

Green energy in thermal systems

The direction of this transition pathway highlights the potential for green energy sources – including CHP and large-scale boilers using natural gas and renewable energy – to reduce the need for grid-based energy supplies, which are still highly reliant on fossil fuels. Renewable energy can be used in various forms for thermal systems, such as capturing solar heat for space or water heating, using biomass for heating, and heat pumps that extract heat or cold from the ground (Casini, 2016). Additionally, buildings can be designed for natural ventilation.

The 13th Five-Year Plan for Renewable Energy Development, issued by the National Development and Reform Commission in 2016, set up the target for all utilisation of renewable energy to reach 730 million tons of coal equivalent (TCE), and all renewable energy power generation capacity reaching 680 GW, by 2020. Among that, the heat collection area of solar thermal utilisation will reach 800 million square metres. By 2020, all forms of renewable energy heating and residential fuel use will replace 150 million TCE of fossil fuels. The 13th Five-Year Plan suggested expanding the use of solar thermal energy in urban and rural areas, actively promoting the development of solar heating technology, and accelerating biomass heating and other non-electricity resources such as geothermal energy for use in buildings.

The National Plan of Heating in Winter in Northern Regions 2017–2021 is a policy that promotes the green energy pathway for northern regions in China, where heating traditionally relied on coal-burning boilers in urban areas and small coal ovens in rural areas. The plan has required '2 + 26' cities to replace coal-fired boilers with natural gas and electricity-powered heaters since 2018. (All the '2 + 26' cities lie within the smog-plagued northern regions, the two cities being Beijing and Tianjin and the other 26 cities including Hebei, Henan, Shandong, and Shanxi Provinces. This plan also set up ambitious targets for clean heating: 50% by 2019 and 70% by 2021. By 2019 clean heating rates in urban areas should reach 90%, fringe areas 70%, and rural areas 40%. By 2021, the clean heating rate in urban areas should reach 100% (removal of all coal-burning boilers less than 30 t/h), fringe areas 80% (removal of coal boilers less than 20 t/h), and rural areas 60%. According to the statistics from the China Ministry of Environmental Protection, over 4.7 million households in 21,516 villages in the inspected area have completed coal-to-gas or coal-to-electricity conversions, 3.94 million of which finished the replacement in 2018 (*Xinhua*, 2018).

However, converting to electric heating will also need to be complemented with energy efficiency actions such as improving building insulation. Many buildings in the peri-urban areas (e.g. fringe areas) have poor insulation performance and the homes are often larger than city apartments. Heating in the peri-urban residential buildings should consume electricity five times more than the city. Additionally, the indoor temperature may not be comfortable for children and the elderly. Greening the heating system needs to consider indoor comfort of inhabitants and their perceived norms, and should also be supplemented with actions from other components of the energy efficiency pathway and the electricity system supplying buildings.

Green electricity

Integrating renewable energy sources into buildings can transform the building sector from an energy consumer to a power generator that supplies energy to the grid (IEA, 2013). Renewable energy generated by the building itself, or from its surroundings, can be used to supply thermal and/or electrical energy where technically feasible and economically viable. For instance, electricity can be generated via photovoltaic (PV) systems and mini wind turbines on the roofs of certain buildings or in neighbourhoods. This could be used for water heating and powering heat pumps, and surplus electricity can be transferred to the power grid.

The 13th Five-Year Plan will help to trigger the long-term development of decentralised renewable energy systems in buildings. By 2020 the building areas that allow for solar hot water installations should have increased to 30 million square metres of construction area, with heating from grounded-thermal supplying 20 million square metres. This national target is enhanced in some cities, for instance in Beijing where buildings with renewable energy should account for 16% of the city's total buildings by 2020.

In China, due to the limited roof access of residential apartments, the main focus of the renewable energy component of the pathway will be on installations in public and commercial buildings. Notice on Related Policies to Further Implement Distributed PV, released by the National Energy Administration in 2014, encouraged the development of various forms of distributed photovoltaic power generation applications and making full use of qualified building roof (including affiliated free site) resources. Furthermore, local governments at all levels have been required to support the installation of PV power generation applications via financial subsidy policies, paying special attention to combing these projects with poverty alleviation to ensure an income increase of ¥3000 per household per year in poor areas.

In addition, micro-grid and renewable energy power projects have also been mentioned in 13th Five-Year Plan for Solar Power Development. It would be useful to construct micro grids for renewable energy in the areas where the distributed renewable energy penetration rate is higher and building conditions suitable for the utilisation of solar resources.

Considering these components of energy efficiency in the building sector and promoting renewable energy in the building sector and the electricity system, the next section will explore the potential barriers to promoting these components as well as the possible unintended negative outcomes of these pathways.

Risk and uncertainty analysis

Stakeholders identified key risks and uncertainties associated with these green building transition pathways and their main components: energy efficiency for new green buildings, energy efficiency for retrofitting existing buildings, and renewable energy. Implementation risks and consequential risks are considered for each building category in the planning, design/construction, and operation phases, as well as for the renewable energy component of the pathway in the installation and operation phases (Figure 10.1).

Implementation risks (barriers)

Stakeholders identified several barriers that may occur before or during the transition process. These include risks from different areas such as incentive policies, regulations, application of energy-efficient technologies, economic factors (market pricing or financing), social factors (behaviour change), and environmental factors. Different stakeholders attribute different levels of concern to these risks (Figure 10.2).

For government stakeholders, the top implementation risks centre on the policies and technologies. They frequently brought up issues with the implementation of energy efficiency certification policies for operations. As of 2014 there are roughly 1500 buildings in China certified by the Three Star System, out of which only 107 had the operation label (Li *et al.*, 2014). In 2015, the operation labels only represented 4.4% of the total certified buildings in Shanghai (MCOHURDM, 2016). The growth of green buildings has largely been fuelled by government policies and targets (Kong *et al.*, 2012), but the green building principle cannot be implemented as expected. The main barriers include both immature green certification and weak policy monitoring and enforcement for the operation of new buildings. The lack of policy incentives for the certificated operation label corresponds with a lack of leadership from the construction industry to meet the standards.

The technological innovations to promote energy saving in the building sector primarily involve two aspects: enclosure structure (doors, windows, roofs, floors, and walls) and equipment systems (heating, cooling, lighting, ventilation, and operation management optimisation). For the enclosure structure parts, stakeholders indicated that the technological innovation and application for enclosure structures, such as envelope materials and roofing insulation, did not consider local demands and energy efficiency goals. Technological innovation should respect the environmental, social, and economic conditions in different local areas otherwise the benefits can be limited as the energy efficiency technologies can increase the costs of construction while having limited impact on energy efficiency. In some buildings, technological applications failed

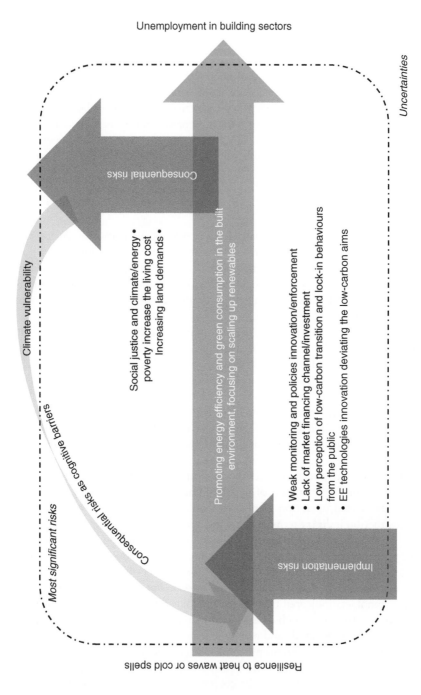

Figure 10.1 Main uncertainties, implementation, and consequential risks associated with a green building transition pathway in China.

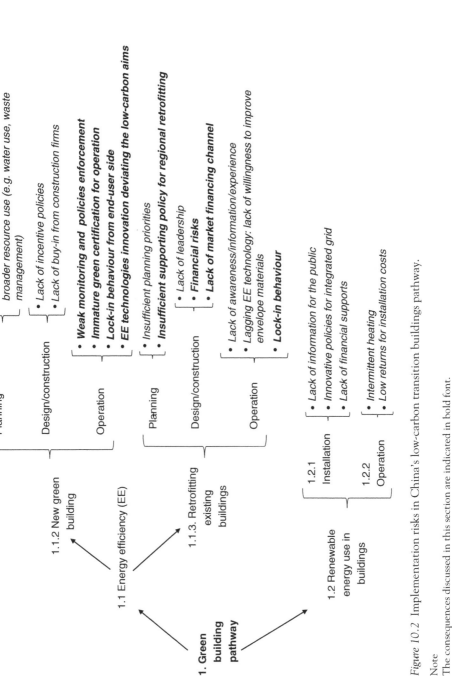

*Implementation risks in italic font

Planning
- *Lack of integrated systemic planning with broader resource use (e.g. water use, waste management)*

Design/construction
- *Lack of incentive policies*
- *Lack of buy-in from construction firms*

Operation
- **Weak monitoring and policies enforcement**
- **Immature green certification for operation**
- **Lock-in behaviour from end-user side**
- **EE technologies innovation deviating the low-carbon aims**

Planning
- *Insufficient planning priorities*
- **Insufficient supporting policy for regional retrofitting**

Design/construction
- *Lack of leadership*
- **Financial risks**
- **Lack of market financing channel**

Operation
- *Lack of awareness/information/experience*
- *Lagging EE technology: lack of willingness to improve envelope materials*
- **Lock-in behaviour**

1.2.1 Installation
- *Lack of information for the public*
- *Innovative policies for integrated grid*
- *Lack of financial supports*

1.2.2 Operation
- *Intermittent heating*
- *Low returns for installation costs*

1.1.2 New green building

1.1 Energy efficiency (EE)

1.1.3. Retrofitting existing buildings

1. Green building pathway

1.2 Renewable energy use in buildings

Figure 10.2 Implementation risks in China's low-carbon transition buildings pathway.

Note

The consequences discussed in this section are indicated in bold font.

to improve energy efficiency because of the social habits of occupants. Also, in some cases, rather than taking a technological approach, similar impacts may be achieved through cheaper non-technical efforts, such as taking advantage of natural environments.

Furthermore, other risks raised by stakeholders from governmental groups focused on behaviour lock-in, not only for the new building but also for retrofitted buildings. New buildings in Shanghai mostly met energy efficiency codes; however, the city has seen a rapid growth in energy consumption of more than 55% in 2016 due to huge economic development and improvements in living standards (especially building heating). Even though the building sector is transitioning towards green buildings, energy saving in the building sector is thought to be diluted by high-energy consumption behaviours. For instance, more and more residents in the Shanghai area are installing central air conditioning for heating and cooling in order to improve indoor comfort. Energy consumption for heating in southern areas is now increasing.

For energy-efficient building retrofitting, the perception and acceptance of households also play very important roles in the process of a low-carbon transition. The 12th Five-Year Plan (from 2011 to 2015) specified that the northern district should carry out 'heating management'. Heating management includes indoor temperature control and installing net metering in existing residential buildings to promote energy saving in an area of more than 400 million square metres. However, the achievements of retrofitting projects have differed among the different climate zones. The target had been almost overachieved in advance in northern areas – 750 million square metres of housing in the northern heating zone has been retrofitted since the end of 2014. However, in contrast with the northern district, retrofits of existing buildings in southern areas has been hindered by several complicated factors. The energy-saving retrofits of existing buildings in these districts are still maintained in the small-scale pilot phases of large public buildings, especially retrofits of external walls' insulation membranes and coatings.

Furthermore, regarding the barriers to renewable energy installations, some stakeholders from the general public had a low acceptance of renewable energy installations. However, subsidies for solar PV installations have been provided by national, municipal, and district level government respectively, providing a subsidy of more than ¥1 per kilowatt hour in Beijing. Under these incentives, the benefits are being recognised by the public. Each solar power installation project is required to operate for at least five years (from 2015 to 2019) in Beijing. Participants in these projects include not only distributed generation companies but also households. Some households involved in this project considered that the cost would be recouped in ten years through governmental subsidies, self-consumption, and selling excess electricity not consumed in the household to the grid.

For government stakeholders, the economic and financing risks are related to energy efficiency of retrofitted buildings. The retrofits of existing buildings mainly refer to buildings built before the year 2000 which cannot meet the current energy efficiency standard. In 2015 there were almost 60 billion square metres of existing buildings in China, and less than 10% are rated as an energy efficient building. Residential energy efficient buildings accounted for 23.1% of existing build area of all types (Lu *et al.*, 2018). Residential buildings in Shanghai have exceeded 600 million square metres, with about 200 million square metres of them built before 2000 and more than 5.5 million households living there. Residential buildings aged over 50 years have reached 14.77 million square metres. Thus, the investment necessary for retrofitting existing building is very high and will rely on the participation of many stakeholder groups from both the market and the public sector. However, stakeholders representing the general public and construction markets perceived low benefits or returns in their expectations for investments in green buildings and showed a lack of willingness to participate in the retrofitting projects.

Stakeholders indicated that changing coal-fired boilers to electric power or natural gas heaters would not lead to reductions in carbon emissions. They also suggested that behaviour change needs to be consider in order to reduce energy consumption and decarbonise the electricity system. For instance, China's smog-plagued northern regions have been replacing coal-fired boilers with natural gas and electricity-powered heaters. Over 4.7 million households in 21,516 villages in the inspected area have completed coal-to-gas or coal-to-electricity conversion, 3.94 million of which finished the switch in 2018, according to the Ministry of Environmental Protection (Bo, 2018). The programme, as discussed earlier, is mainly implemented in the rural area of northern China, including the 2 + 26 cities. Experts warned that if coal-fired boilers are simply replaced with electricity-powered heaters, there may be no energy saving because over 70% of electricity is generated from coal and the efficiency of coal-burning power generation is less than 40%. Better alternatives would include heat pumps. Clean heating with gas, electricity, clean coal, and renewable energy including biomass, wind, solar, etc. cover about 34% of the total built area now; however, completing this transition is a great challenge for China, especially in rural areas. Beijing and Tianjin have almost completed their transitions, but other cities may have difficulties.

Consequential risks (negative impacts) and uncertainties

A low-carbon transition in the building sector may bring unintended negative impacts on economic efficiency, social development, technological innovation, and environmental sustainability. All are risks related to energy efficiency and renewable energy components of the green building pathway (Figure 10.3). There are fewer consequential risks identified compared to implementation risks, since many of the negative impacts are not yet known for these relatively new energy initiatives in the building sector.

Stakeholder concern is mainly focused on potential negative impacts on economic and social factors, along with environmental conditions. Stakeholders from all the different groups agreed that the main consequential risks result from

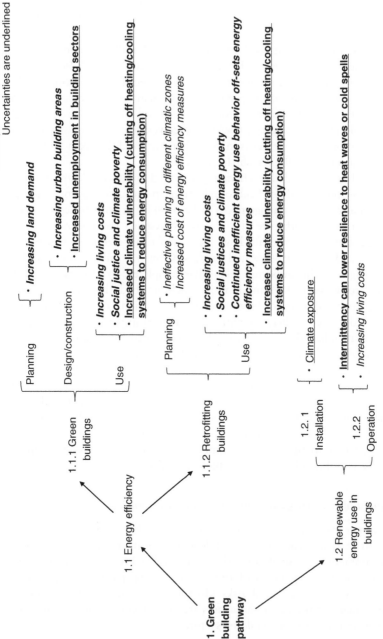

*Consequential risks are in italic font
Uncertainties are underlined

Figure 10.3 Consequential risks in China's low-carbon transition buildings pathway.

Note

The consequences discussed in this section are indicated in italic font. The uncertainties discussed are underlined.

increasing land demands and increasing living costs associated with increasing green buildings to cater for the growing urban population. Even though buildings are becoming greener, increasing urbanisation may trigger an influx of energy consumption in the building sector (Kim and Jeong, 2018). Furthermore, the construction cost of green buildings can drive up the price of real estate, which forces urban residents to suffer from increasing living costs and may result in social inequality or climate poverty such as 'green gentrification' in the process of urban renewal. Furthermore, the cost of retrofitting of around US$30 per square metre would be subsided by the central and local governmental funding in northern areas. Without abundant policy support, residents in southern areas may need to pay for part of this cost due to lack of subsidies and support from local governments, which may also increase their living costs.

However, regarding the risks related to social justice and climate poverty, some stakeholders and experts have suggested that proper institutional arrangements and social welfare policies, such as public rental housing for urban poverty, could offset the negative impacts of the low-carbon transition pathway.

Energy efficiency efforts generally help to improve thermal insulation through the transformation of the external envelope, reducing power consumption by air conditioning use in summer and electric heat in winter. However, residents in southern areas have entrenched behaviours that reduce the impact of energy efficiency behaviour. For instance, in Shanghai and other southern areas with a humid climate, residents often keep their window open in all seasons. In this case, retrofits of thermal insulation have very limited impact.

The main uncertainties raised by the stakeholders derived from the environmental cluster. For instance, intermittency of renewable energy supply for heating and cooling can lower resilience to heat waves or cold spells. Stakeholders also raised climate exposure for green building technologies and renewable energy installations, and increasing climate vulnerability (e.g. cutting off heating/cooling systems to reduce energy consumption). In addition, some stakeholders from the construction markets mentioned there was an uncertain shift for employment in the building sector due to shrinking profits in the construction industries, resulting from the higher costs of innovation, prefabricated technologies, and applying Building Information Modelling (BIM) systems.

Both implementation and consequential risks in the green building transition pathways highlight some of the challenges linked to the application of energy efficiency measures and new green technologies that could have adverse social-economic as well as environmental impacts. The next section synthesises our findings.

Conclusions

The proposed low-carbon transition pathway in China's green building sector emphasises the present policy ambition to scale up green buildings through increasing energy efficiency and promoting renewable energy in buildings. The pathway considers the existing energy policies and urbanisation strategies in

China, and explores the risks and uncertainties in the green building transition pathway, drawing on examples from cities in northern and southern climatic regions.

Implementation risks identified differed according to stakeholders' perspectives, from governments, construction firms, public residents, and the area experts. That said that implementation risks are mainly derived from four clusters including: policies, technology innovation, financing, and energy consumption behaviour. In the low-carbon building pathways, the stakeholders highlighted the lack of policies for green building operation and financial policies to incentivise renewable-energy use as a priority risk. However, the locked-in behaviour of building end users and technological innovation would be the main barriers in the future transitions of green new buildings and retrofitting buildings.

Stakeholders also mentioned that energy-efficient building technologies for the enclosure structure and equipment systems did not often consider local needs to control indoor temperature based on the local climate zone. Energy efficiency measures in northern regions, which has relatively dry cold and hot conditions, cannot be directly replicated in the more humid southern regions where building require additional ventilation.

Aside from different user needs, awareness of cost–benefit analysis for green buildings is still weak in the design of Chinese buildings. Besides this, stakeholders from the government and public groups considered that the efficiency of equipment systems still suffer from a lack of innovation. This especially applies to passive energy-saving technology and heating (air conditioning) systems based on the full use of renewable energy such as solar and wind (Zhang, Kang, and Jin, 2018). For instance, most buildings still rely on air conditioning for cooling, but buildings can be more effectively designed to keep cool by natural winds. Stakeholders indicated that the technological innovation in equipment systems (heating or cooling) needs to move forward in this direction.

Furthermore, the consequential risks considered by stakeholders mainly focused on economic and environmental impacts from the low-carbon transition process. They especially highlighted how the potential cost burdens of green building may play a negative role on social and economic development, such as increasing unemployment and inequality. In addition, stakeholders were concerned that green building developments that do not allow user control could also reduce the environmental benefits. The scaling up of building development could place pressure on land use in the transition process.

Regarding uncertainties, stakeholders argued more about the impact of increasing climate change and extreme climate events such as heat waves and floods on urban environments. Even though we still cannot be sure if climate shocks will bring negative impacts on a low-carbon transition in the building sector, stakeholders still mentioned the potential possibility of increasing exposure of renewable energy installations under climate shock. There were also some uncertainties which prevented market participation, due to perceived investment risks in green buildings and the low acceptance and awareness of green buildings.

Even though mixed quantitative and qualitative methods have been applied in this research, the outcomes have been limited by missing data for some of the target areas, such as Shanghai and other cities in southern areas. This may create barriers to the comparative analysis of transition risks and uncertainties among different climate zones. Currently there are overarching energy efficiency objectives at the national level, and some city-level energy efficiency initiatives in Beijing and Shanghai, but limited guidance on energy efficiency actions in different climate zones. The preferences and indoor behaviour of stakeholders are not well understood at the higher policy levels, thus a 'blanket' approach is taken to address energy efficiency issues in the same manner in all regions. More time needs to be taken to understand buildings' end users in different climatic zones and to design policies that reflect the different conditions and behaviour in order to maximise the impact of energy efficiency measures. Overall, this research has listed and discussed certain implemental and consequential risks in the green transition, but there is still uncertainty about what the priority risks are that need to be addressed in the present and future green transition.

There were some limitations to this study, as the risk and uncertainties assessment assumed that there were no drastic changes in the political environment and regulations. Stakeholders did not make reference to the impacts of political factors on green transitions. This could be due to the fact that China's political system has been stable over past decades and there has been strong policy support from the central government to push forward decarbonisation efforts. However, there are still uncertainties linked to how provincial government may implement national policies considering the varied socio-economic conditions and institutional capacity across provinces.

Moving forward, there are synergies linked with promoting energy efficiency pathways and addressing broader environmental initiatives and challenges. For instance, energy efficiency efforts can be promoted to reduce air pollution in Chinese cities. Air pollution is linked to emissions from energy generation and consumption for both private and industrial uses. Decarbonising the energy systems and improving energy efficiency in the energy system, as well as the transport sector, will therefore also lead to reducing air pollutants in cities. Additionally, China's Nationally Determined Contributions for the Paris Agreement specifically promote a 'low-carbon way of life' through behavioural changes that are directed at society and public institutions. Changes can occur through educating citizens to have more green, sustainable, and healthy ways of life, along with promoting low-carbon consumption patterns. Public institutions, including government buildings, hospitals, higher education campuses, and military buildings, are to take lead on promoting low-carbon initiatives and reducing waste.

This chapter introduces key elements to understand China's low-carbon transition trajectory. However, there are other related but hidden drivers of China's low-carbon transition, such as technology transfers from other advanced economies, which in turn created a wave of technological start-ups. For instance, local governments and fierce competition among regions created business-friendly environments driving the rise of the PV industry.

References

Bo, X. (2018). Workers help warm rural Beijing after coal-to-electricity conversion. Available at: www.xinhuanet.com/english/2018-01/02/c_136866891.htm.

Building Energy Research Center of Tsinghua University (BERC) (2015). *China Building Energy Use 2016*. China Architecture & Building Press.

Casini, M. (2016). Designing the third millennium's buildings. *Smart Buildings: Advanced Materials and Nanotechnology to Improve Energy-Efficiency and Environmental Performance*, 3–54.

Chen, W.K., Luo, F. (2008). Research on building energy consumption based on whole life cycle theory. *Building Science*.

Chen, C. (2018). *The Perspectives of Green Building Development in China (2020)*. Available at: www.qianzhan.com/analyst/detail/220/180423-d50ad8bc.html.

Fastenrath, S., Braun, B. (2018). Ambivalent urban sustainability transitions: insights from Brisbane's building sector. *Journal of Cleaner Production* 176, 581–589.

Geng, Y., Dong, H., Xue, B., Fu, J. (2012). An overview of Chinese green building standards. *Sustainable Development* 20(3), 211–221.

IEA (International Energy Agency) (2013). *Transition to sustainable buildings: strategies and opportunities to 2050*. IEA.

IEA (International Energy Agency) and Tsinghua University (2015). *Building Energy use in China: Transforming construction and influencing consumption to 2050*. OECD/IEA.

Jiang, R., Li, R. (2017). Decomposition and Decoupling Analysis of Life-Cycle Carbon Emission in China's Building Sector. *Sustainability* 9(5), 793.

Kim, T.H., Jeong, Y.S. (2018). Analysis of Energy-Related Greenhouse Gas Emission in the Korea's Building Sector: Use National Energy Statistics. *Energies*, 11(4), 855.

Kong, X., Lu, S., Wu, Y. (2012). A review of building energy efficiency in China during 'eleventh five-year plan' period. *Energy Policy*, 41624–41635.

Li, Y., Yang, L., He, B., Zhao, D. (2014). Green building in China: needs great promotion. *Sustainable Cities & Society* 11(complete), 1–6.

Li, Y., Ren, J., Jing, Z., Lu, J., Ye, Q., Lv, Z. (2017). The existing building sustainable retrofit in China: a review and case study. *Procedia Engineering* 205, 3638–3645.

Lu, Y., Cui, P., Li, D. (2018). Which activities contribute most to building energy consumption in China? A hybrid LMDI decomposition analysis from year 2007 to 2015. *Energy and Buildings* 165, 259–269.

MCOHURDM (2016). *Shanghai Green Building Development Report*. Shanghai Municipal Commission of Housing, Urban–Rural Development and Management.

MOHURD (Ministry of Housing and Urban–Rural Development) (2017). *Action Plan of 'Thirteenth Five-Year Plan'*. Assembly Building.

NBS (National Bureau of Statistics of China) (2016a). *Total population 2016*. http://data.stats.gov.cn/english/adv.htm?m=advquery&cn=C01.

NBS (National Bureau of Statistics of China) (2016b). *Total Energy and Coal Consumption 2016*. http://data.stats.gov.cn/english/adv.htm?m=advquery&cn=C01.

Pachauri, R.K., Allen, M.R., Barros, V.R., Broome, J., Cramer, W., Christ, R., … Dubash, N.K. (2014). *Climate change 2014: synthesis report. Contribution of Working Groups I, II and III to the fifth assessment report of the Intergovernmental Panel on Climate Change* (p. 151). IPCC.

Tsinghua University (2016). *Annual Report on China Energy Efficiency 2016*. China Building Industry Press, Beijing.

UNEP (2011). Climate change and pops: predicting the impacts. Report of the UNEP/ AMAP expert group. *Arctic Monitoring & Assessment Program.*

Wei, W., He, L.Y. (2017). China building energy consumption: definitions and measures from an operational perspective. *Energies* 10(5), 582.

World Bank Group (2018). *China Overview.* Available at: www.worldbank.org/en/country/china/overview.

Xinhua (2018). Workers help warm rural Beijing after coal-to-electricity conversion. www.xinhuanet.com/english/2018-01/02/c_136866891.htm.

Yu, S., Evans, M., Shi, Q. (2015). Analysis of the Chinese market for building energy efficiency. *Current Politics & Economics of Northern & Western Asia* 24.

Yuan, X., Zhang, X., Liang, J., Wang, Q., Zuo, J. (2017). The development of building energy conservation in China: a review and critical assessment from the perspective of policy and institutional system. *Sustainability* 9(9), 1654.

Zhang, Y., Kang, J., Jin, H. (2018). A review of green building development in China from the perspective of energy saving. *Energies*, 11(2), 334.

11 Greece

From near-term actions to long-term pathways – risks and uncertainties associated with the national energy efficiency framework

Alexandros Nikas, Nikolaos Gkonis, Aikaterini Forouli, Eleftherios Siskos, Apostolos Arsenopoulos, Aikaterini Papapostolou, Eleni Kanellou, Charikleia Karakosta, and Haris Doukas

Introduction

At both the community and the member state level, the European Union (EU) has committed to actions focusing on enhancing energy efficiency in pursuit of mitigating greenhouse gas (GHG) emissions and enhancing security of energy supply and socio-economic sustainability. Drawing from its respective national commitments as well as the need to respond to the European and global efforts on climate change, Greece too has recently been striving to design and implement an effective and sustainable energy efficiency policy framework.

These efforts have been a dynamic learning process through which the policy framework is redesigned along the way. Such a framework encompasses policy instruments, measures, and interventions, including financial incentives and tax breaks, in the energy efficiency area. These actions primarily regard the built environment but to some extent are also focused on the energy efficiency of the transport sector. At the same time, since energy efficiency programmes in the building sector also include initiatives for residential micro-generation from renewable sources, this framework has implications for energy policy and is synergistic with further diffusion of renewables. So far, the policy framework has been oriented towards financial incentives for the diffusion of renewable energy as well as a building renovation strategy across all scales: residential, public, and private.

In respect to progress in implementing the Energy Efficiency Directive, Greece, among other countries (Fawcett, Rosenow, and Bertoldi, 2017), appears to be on track to achieving its energy efficiency goals of 20%; it does, however, lag behind its goals in respect to achieving new savings of 1.5% of the annual energy sales to final consumers every year until 2020 (Nikas, Ntanos, and Doukas, n.d.). On the renewable energy front, the country appears not to be on track either (Capros *et al.*, 2016), with respect to achieving the Community-set target of 18% (Directive 2009/28/EC), which has since been amended to 20% in the more ambitious Greek legislation (Law 3851/2010). This has implications for energy efficiency in buildings since various energy-efficient actions are focused on electricity generation from building-integrated photovoltaics (PVs). According to the European Commission's annual progress reports, there is a significant divergence between the intermediate targets and the actual achieved savings, resulting from the inadequate implementation of poorly designed previously proposed actions. In fact, the latest official Greek report acknowledges the socio-technical infeasibility to meet the predefined objectives, with an expected gap of 35% (1176 kTOE) of the overall energy savings target by 2020 (Ministry of Environment and Energy, 2017).

It is noteworthy that, despite the reported progress, there exists large potential for improving energy efficiency in the country. The building sector, in particular, has significant room for decarbonisation as 25–30% of the final energy is consumed in the residential sector (Centre for Renewable Energy Sources and Saving, 2015). Considering the current building stock has predominantly poor energy performance, with about six out of ten buildings having been constructed before 1980 (Hellenic Statistical Authority, 2011), it is consequently in need of immediate renovations. The same can be said of the transport sector, with a share in final energy consumption of about 35% during the last decade, according to the International Energy Agency, and a 15- to 20-year old transport fleet (Tsita and Pilavachi, 2017).

Given this potential, it is evident that current difficulties in delivering on the national energy-efficiency commitments can, at least in part, be attributed to the underlying risks and uncertainties that essentially did not allow for the successful implementation of the policy framework. At the same time, there has been little provision for potential negative outcomes of relevant policies, thereby allowing adverse consequences to manifest. For example, despite their positive impacts (HELAPCO, 2016), policies aimed at further developing the solar power sector have had detrimental economic side effects (Tselepis, 2015) and a series of amending regulatory efforts have created an uncertain investment environment in the energy front.

Until recently, the national policy framework on energy efficiency has comprised actions, incentives, and interventions, the design of which had been based on quantitative modelling exercises. By looking at the inadequacy of this framework, as reported in the latest national energy efficiency action plan (Ministry of Environment and Energy, 2017), it seems that scientific and policymaking processes have ignored a number of implementation risks and uncertainties. They have also failed to foresee the manifestation of negative consequences resulting from the policy framework's implementation. These risks and uncertainties, if overlooked, could jeopardise both the national energy efficiency framework towards 2030 and overall efforts towards long-term decarbonisation. Consequently, the resulting failure to accomplish the near-term energy efficiency national goals raises questions, not only about the country's capacity to realise long-term decarbonisation visions, but also about the existence of such visions. In this context, it is necessary to carefully examine the risks and uncertainties before reviewing and redesigning the strategies to contribute to a sustainable transition pathway in both the near and the longer term.

The aim of this chapter, therefore, is to discuss underlying uncertainties and risks threatening the transition to an energy-efficient Greek economy. We focus mainly on the built environment yet with some implications for transportation and power generation. Such a transition prospectively looks at a future Greek building stock consisting of near-zero-energy buildings in the residential, private, and public sectors, with autonomy-oriented infrastructure, based on decentralised power (micro-)generation and electricity storage technologies. Acknowledging their capacity to help identify the strengths and weaknesses of current and future policy frameworks (Nikas *et al.*, 2017), this discussion is carried out from a stakeholders' point of view, focusing on the risk and uncertainty dimension. This includes the potential negative effects of these strategies as well as the perceived barriers that may hinder the successful uptake of energy efficiency measures, both in the short and longer term.

The following section features a discussion of the context of the study, highlighting the uncertain environment characterising the Greek case. Then, the immediate implementation risks that jeopardise effective uptake of near-term policy instruments are discussed, followed by an analysis of how stakeholders perceive both implementation and consequential risks associated with a transition over the longer term. Finally, the conclusions of the chapter are presented, in addition to the limitations and future prospects of this study.

Research process and methods

For the purposes of exploring policy strategies for meeting near-term energy efficiency targets, the most critical implementation risks were identified in a meeting with experts from the Ministry of Environment and Energy, during which the above uncertainties were extensively discussed. Subsequently, a number of policy instruments found in the latest National Energy Efficiency Action Plans (NEEAPs) of Greece were evaluated by means of a multi-criteria group decision-making (MCGDM) methodology. In a series of interviews, seven stakeholders from the ministry were asked to assess the vulnerability of each of the policy measures against the identified risks and, based on their input, the MACE-DSS tool (Nikas *et al.*, 2018) was used to rank the policy instruments in terms of risk vulnerability. This approach, (similarly to Doukas, Karakosta, and Psarras, 2009), gave us a good overview of how key experts from the ministry view the proposed measures, in respect to the actual capacity of the country to implement them. The analysis was used to complement a portfolio analysis (PA) approach for determining the optimal near-term policy mix, the findings of which are also discussed.

When considering a low-carbon transition over the longer term, experts are able to identify a different, richer set of risks. From this perspective, a policy strategy aiming to promote a pathway is not only prone to implementation risks but may also lead to foreseeable negative outcomes (consequential risks). Using the same MCGDM methodology, a different, more diverse group of ten stakeholders was interviewed in order to elicit their knowledge and assess the risks themselves, this time against specific evaluation criteria. The engaged expert group comprised of three policymakers from the Ministry of Environment and Energy (also included in the first stakeholder engagement process), two researchers, one representative of the financial sector, one representative of electric utilities and regulators, one member of associations involved in the provision of GHG emissions, and two technology suppliers (including manufacturers and importers). This process gave us a good overview of how different experts perceive the underlying implementation and consequential risks, providing us with insights into which of these risks appear to be more critical for the successful promotion of a sustainable transition pathway. Furthermore, this analysis was used to complement a fuzzy cognitive mapping (FCM) approach (Nikas and Doukas, 2016) for determining, from the experts' perspective, the policy focus that appears to be most robust, the findings of which are also discussed.

Context: an uncertain environment

The policy framework in Greece is uncertain per se: an overview of the past pathway indicates that, as far as energy efficiency is concerned, the country has seen significant delays in adopting European directives into the national policy framework. After the early determination of thermal insulation requirements for buildings in 1979, the first notable policy in the country was the late adoption of the 1993 EU directive on limiting carbon emissions by improving energy efficiency (in 1998). Respectively, National Law 3661/2008 brought Directive 2002/91/EC into effect by defining minimum energy performance requirements, introducing energy performance certificates and referring to qualified and accredited energy inspectors. It was amended in 2010 so as to provide for the gradual implementation of energy management systems in all public buildings. Although the adoption of the Regulation on the Energy Performance of Buildings (KENAK) in 2010 is considered to have set a milestone in national energy efficiency policy – and has since been updated based on European regulatory advancements – it was the belated Law 4342/2015 that adopted the Union's 2020 objectives (Directive 2012/27/EU). This record of significant delays in adopting European directives into the national policy framework can itself be considered as an uncertainty for future developments in both energy and climate policy.

On the micro-generation front, which could prove instrumental in accomplishing renewable energy targets and is included in energy efficiency improvement programmes, large investments took place in a period of strong fiscal incentives for boosting solar power development (2008–2012). This eventually led to an excessive burden on the operator. Due to the ever-growing deficit, the government proceeded with a 12.5% cut in the support scheme for solar installations commissioned after February 2012 and a larger (44%) cut on feed-in tariff (FIT) rates for solar PV plants installed after January 2013. In fact, it was only the attractive FIT contracts still in effect that sustained the momentum of solar power growth in 2013, despite the economic recession and the unfavourable changes to the incentive system. The latter, in combination with a freeze on the receipt and processing of new solar power investment applications between August 2012 and April 2014 as well as an unexpected shift towards new lignite-fired plants, constitute the background to an ever-changing regulatory environment and the consequent caution and mistrust in the policy framework.

Moreover, public perception of climate change and the urgency of the need to mitigate its impacts also appears to change in a period of economic crisis. In particular, Greek citizens seem to perceive the problems associated with climate change as a matter of priority. The vast majority of the Greek people are aware of environmental problems and, at the break of the recession, considered climate change as the ultimate global problem (European Commission, 2008). However, five years later, trends were reversed as poverty and the economic situation dramatically gained on climate change in terms of prioritisation (European Commission, 2013). At the same time, citizens now appear to have very low confidence in authorities and big enterprises with regard to their capacity and willingness to deal with climate change (Papoulis *et al.*, 2015). Uncertain societal cohesion adds to the already volatile context of Greek efforts towards energy efficiency and decarbonisation, since a lack of public acceptance could potentially halt the introduction and diffusion of technically and economically feasible technological options, as well as the successful implementation of otherwise prominent policy instruments. Especially with regard to enhancing energy efficiency in the residential and commercial sectors, many of the measures considered depend on behavioural change and rely heavily on initiatives that decision makers, at the small scale, are free to implement or disregard.

Last but certainly not least, Greece also faces challenges and respective uncertainties in the political and economic areas, with impacts on the energy front (Doukas *et al.*, 2014). It is a country hit hard by the 2008 crisis, which led to a cumulative output loss of 30% since the beginning of the recession, 45% of the population living below the poverty line (when considering a poverty threshold anchored to 2008 in real terms), and consequent inequality and decline in average living standards. This has implications for long-lasting societal consequences (Kaplanoglou and Rapanos, 2018). As a result, Greece also went through significant political instability and is still mired in a crisis of political representation (Stavrakakis and Katsambekis, 2018). A by-product of this instability was an unexpected approval of the construction of new lignite-fired power plants, namely Ptolemaida V and Meliti II, of combined capacity of more than 1 GW (Simoglou *et al.*, 2018), giving rise to doubts about how determined the Greek government is to actually deliver on its energy and climate commitments.

Based on these conditions, many critical questions emerge. Will Greece be politically and financially capable of incentivising near-term advancements and long-term transitions? Even if this is the case, will citizens have the economic capacity and adequate trust in the stability of the policy framework to take advantage of the support mechanisms? It is evident that these uncertainties can foster the manifestation of severe implementation risks to any policy framework, or allow for negative consequences of the latter to realise.

Challenges in the short to medium term

In light of the country's near-term energy efficiency commitments, experts from the Greek Ministry of Environment and Energy were mostly interested in exogenous factors that pose challenges to the successful implementation of a new policy framework, i.e. implementation risks. This can mainly be attributed to the fact that the time horizon discussed, i.e. 3–10 years, was not perceived to be long enough to accommodate a reliable evaluation of negative consequences.

Ministry experts agreed that a significant challenge for any strategy oriented on enhancing energy efficiency in Greece lies in public perception. Citizens must, first and foremost, comprehend the benefits of sustainable energy use; this has not been done successfully in the past, and this type of limited awareness threatens the implementation of respective measures in the residential sector, which covers about 79% of the national building stock (Gaglia *et al.*, 2017). There appears to be a cognitive trade-off between the benefits of adopting a sustainable energy behaviour and a household's trust in a consistent and rewarding regulatory environment, i.e. in the government's ability to maintain a stable regulatory framework that compensates citizens for taking up energy efficiency measures. Despite this, compared to other possible actions, energy-efficient housing-related subsidies and policy instruments (and, therefore, any initiative aiming at the residential sector) currently present the highest level of satisfaction, and are thus not considered a priority for the state to do more (Zerva *et al.*, 2018).

Lessons from failures of the past, such as the first 'Save Energy at Home' financial mechanism for residences, must be used to optimally design future financial support schemes. So far, financial institutions, with an established role as the broker, have not properly supported the processes, adding to the already overburdened bureaucracy associated with applications for such programmes. Due to the adverse economic environment, the banking sector is also seen to be inadequate in promoting energy-efficiency-targeted renovations from a financial capacity point of view.

Along the same lines, experts collectively pointed out the current difficulty for the vast majority of the population of investing in building renovations, let alone in constructing near-zero-energy buildings. The construction sector has almost completely shut down due to the ongoing recession, as well as the severe tax burdens imposed in the context of austerity. In fact, these burdens appear to be among the few constant regulatory priorities throughout this period, which are described by experts as politically unstable, to a point that they can severely jeopardise the consistent implementation of the energy efficiency policy framework. Some experts, in particular, suggested that policymaking per se has lately been carried out recklessly, without looking far ahead, and almost exclusively in compliance with the austerity-driven financial commitments of the country. This is an interesting suggestion, coming from the policymaking stakeholder group. Eventually, this gives rise to the risk of a regulatory framework that is too demanding in relation to the maturity of the market and does not successfully attract investments.

Most measures in the past focused on energy efficiency in public buildings, for both central government and local authorities and municipalities. Although this is not generally perceived as an example of poor prioritisation, the risk of not managing to align the interests of local governments with the national commitments and directions should be acknowledged. A recent example can be found in the issuing of Energy Performance Certificates (EPCs), where no coordination actions among stakeholders and/or between the central and local governments reportedly took place to ensure coherent and transparent implementation (Spyridaki *et al.*, 2016).

Ultimately, in order to take proper mitigation action, it is important to understand the risks associated with the individual policy instruments considered by the ministry for its previous and new Energy Efficiency Action Plans (Ministry of Environment and Energy, 2014, 2017) – see Table 11.1. These included: the 'Save Energy at Home II' financial programme for incentivising energy upgrade actions in the broader residential sector; the energy upgrade programme for public buildings, with regard to building shells, lighting, and building energy management systems (BEMS); energy efficiency and demonstration projects for small-medium enterprises (SMEs); the implementation of an energy management system in the broader public building sector, in accordance with the ISO 50001 standard; an interest rate subsidy for funding energy improvement actions in commercial buildings, through energy service companies (ESCOs); a wide-scale deployment of smart metering systems; a series of large-scale environmental infrastructure projects and other interventions at the national level, as part of the operational programme Environment and Sustainable Development (OPESD); the offset of fines on illegal residential buildings with costs for services, tasks, and materials used for the energy upgrade of these buildings; the definition of energy management duties and the implementation of action plans in municipal buildings; the extension of the district heating networks from the integrated expansion of the Ptolemaida and Amyntaio network, as well as the planned expansions of the Florina and Kozani district heating networks; the replacement of old public and private light trucks; incentives for the replacement of old private passenger vehicles; a retrofitting programme for street lighting; financial support for retrofitting pumping systems at the municipal level; and dissemination activities for behavioural change oriented on the benefits of EPCs.

Eight specific implementation risks are associated with one, several, or all of the policy instruments. These included: the inability of local governments to align their priorities with the obligations of the central government; political instability; complex bureaucratic processes; demanding regulatory framework in respect to market maturity; the inadequate banking sector; limited societal

Table 11.1 Most critical implementation risks for each of the considered near-term policy instruments, according to ministry experts

Policy instrument	Primary implementation risk	Secondary implementation risk
'Save Energy at Home II' financial programme	Complex bureaucratic processes	Inadequate banking sector
Energy upgrade of public buildings	Difficulties in aligning local authorities with obligations of the central government	Inadequate banking sector
Energy efficiency and demonstration projects in SMEs	Complex bureaucratic processes	Inadequate banking sector
Implementation of a BEMS in the broader public building sector, based on ISO 50001	Difficulties in aligning local authorities with obligations of the central government	Inexperienced personnel and lack of technical skills
Incentives for ESCO – commercial building owner partnerships	Inadequate banking sector	Unfavourable market conditions
Large-scale deployment of smart metering systems	Complex bureaucratic processes	Unfavourable market conditions
OPESD actions	Complex bureaucratic processes	Difficulties in aligning local authorities with obligations of the central government
Offset of fines with costs for energy upgrade actions on illegal buildings	Complex bureaucratic processes	Limited societal acceptance
Energy management duties and action plan implementation in municipal buildings	Difficulties in aligning local authorities with obligations of the central government	Inexperienced personnel and lack of technical skills
Extension of the district heating network	Complex bureaucratic processes	Demanding regulatory framework in respect to market maturity
Replacement of old public and private light trucks	Complex bureaucratic processes	Inadequate banking sector
Incentives for replacing old private passenger vehicles	Complex bureaucratic processes	Inadequate banking sector
Retrofitting street lighting	Complex bureaucratic processes	Inadequate banking sector
Retrofitting pump stations	Complex bureaucratic processes	Inadequate banking sector
Promotion of EPCs and behavioural change	Difficulties in aligning local authorities with obligations of the central government	Inexperienced personnel and lack of technical skills

acceptance; inexperienced personnel and lack of technical skills; and unfavourable market conditions. They can be distinguished as primary and secondary risks.

The most critical risk appears to lie in the complexity of bureaucratic procedures necessary to approve, apply for, and secure funding for or implement most of the considered policy instruments. The most interesting observation is that this risk appears to be the most critical among all eight risks not only for the policies targeted at the residential and commercial sectors, but also for some of those involving responsibilities of and actions by the central government and local authorities. It is also noteworthy that, especially with regard to the latter, experts appear to stress the importance of transparent and effective communication and co-ordination between the central and local governments for almost all policy instruments acting at the municipal level.

Risks of a financial nature, namely the inadequacies of the involved financial institutions (banks) to support renovation actions and the unfavourable economic environment, can also prove critical to successfully implementing policy instruments targeted at the residential and commercial sectors. Inexperienced personnel and poor technical skills required at all stages of renovation work is another crucial risk. Interestingly, this was considered not only to potentially hinder the successful installation and management of BEMS in public buildings, but also to constitute a barrier to the successful communication of the benefits of energy efficiency actions to citizens.

Among the remaining implementation risks, political instability was not considered among the most critical for any of the 15 measures. This can in part be attributed to the fact that any turbulence in the political scene may be considered to be reflected in societal acceptance, given the short time horizon and the equally limited time and capacity to drastically change the regulatory framework.

The most 'at risk' policy instruments appear to be the offsetting of fines on illegal buildings with energy efficiency-related interventions, and the replacement of old private passenger vehicles. These are closely followed by: the implementation of the OPESD programme; the energy upgrade of buildings in the broader (central and municipal) public sector; and the appointment of energy managers in public buildings. In other words, stakeholders appeared to view policies targeted at the residential sector as more vulnerable to existing implementation risks. This is mainly because of the complexity of the associated bureaucratic processes and the adverse economic conditions, followed by policy instruments targeting buildings of the broader public sector and especially those at the municipal level (Doukas and Nikas, 2019).

This is also why only one of these highly vulnerable instruments among the measures, namely appointing energy managers at the municipal level, appears to be beneficial to achieving near-term energy savings, in an approach aiming to both maximise cost effectiveness and minimise risk. Aside from this risky instrument, the other two most beneficial instruments were found to be the 'Save Energy at Home II' programme and the energy efficiency and demonstration projects in SMEs. It is also worth to note that a policy portfolio mainly comprising these three instruments targets all sectors (residential, commercial, and public). Other instruments found to appear in the most robust portfolios include the deployment of smart metering systems and the promotion of EPC benefits at the residential sector, as well as the retrofits of street lighting and pump stations at the municipal level (Forouli *et al.*, 2019). Regarding the wide-scale deployment of smart meters, although not as cost, energy, or risk efficient as other instruments considered, they were included in the resulting optimal policy mixes as a fixed budget has been secured by the Hellenic Electricity Distribution Network Operator exclusively for this instrument.

Challenges, opportunities and risks in the longer term

All stakeholders acknowledged that the Greek building sector is the sector with the largest potential for easy-to-implement improvements in energy consumption, along with power generation. For a country with Greece's potential for solar and thermal solar system exports, there are many opportunities in promoting solar heating and cooling systems. Rational energy use is also based on the installation and management of optimised BEMSs, replacement of old devices and motors of low energy performance, investments in building shells and insulation, diffusion of other renewable energy sources in the built environment (e.g. geothermal power), and wide-scale development of building-integrated photovoltaics, which constitute both available and efficient technological solutions but are expensive. Especially in the Greek islands, where demand varies significantly throughout the year, use of large solar power storage batteries could potentially help overcome the current challenges in promoting sustainable energy without heavy investments in linking the mainland and non-interconnected networks or new power plants.

This sectoral preference, however, is concerned not only with energy efficiency but also with the vision of the overall Greek low-carbon transition in general. This can be partly explained by the perceived infrastructural challenges associated with the transformation of the transport sector; the limited financial capacity to invest in decarbonising practices in either an otherwise small heavy industrial sector or a practically non-existent light industrial sector, with negligible potential for emission reductions; and the limited interest from those involved in the agricultural sector.

Implementation risks

Stakeholders agreed on ten major implementation risks with regard to existing barriers posing direct or indirect threats to the successful design, funding, and implementation of a sustainable and effective energy efficiency policy framework. On the social axis, stakeholders considered that societal participation is the major barrier to achieving an energy-efficient economy; however, they disassociated the *lack of public awareness* from *distrust of government or institutions and respective societal opposition*. From a political perspective, instability in the

Greek political scene appears to be a severe implementation risk. This is primarily seen as a short-term risk; however, it is also associated with longer-term uncertainties, such as the economic environment. This instability is also reflected in recent high abstention levels, frequent movements of members of parliament among elected parties, an inability to form single-party governments, as well as the weak ruling of coalition governments and a record of snap elections. In the longer run, the main concern of stakeholders appears to be the *poor prioritisation of climatic change and action, and political inertia.* Amidst the ongoing economic crisis, and in light of the respective national commitments, all political parties' priorities are perceived to be mainly of socio-economic nature and to orient on addressing the recession, rather than climate change mitigation and sustainable energy use. This is why policymaking in Greece is believed to focus on a short-term perspective without a clear and solid strategy, mainly driven by commitments to the austerity-oriented memoranda.

With regard to the Greek economy, all stakeholders are significantly concerned with the adverse economic environment in Greece and the respective *lack of financial capacity.* This began with the difficulties in overcoming what appeared to be a short-term recession at a global scale but is still in effect and is estimated to have long-lasting economic implications for the average Greek household. This concern refers not only to the capacity of the state to incentivise a low-carbon transition in the longer run, but also to the capacity of citizens to make use of available incentives and invest in energy upgrades for their dwellings. Additionally, this particular implementation risk is even more complicated in the context of the built environment since the economic situation is also a key driving force for the construction sector.

When looking at the transition pathway itself, rather than a number of policy strategies designed specifically with the aim of promoting such a pathway, stakeholders are also concerned with the possibility of a general *lack of economic incentives, subsidies, and tax breaks.* The latter, however, is deemed to be of lower importance since incentives and other financial mechanisms are considered primarily dependent on the will, determination, and financial capacity of the government, rather than on a multitude of exogenous factors. It is noteworthy that the three engaged policymakers almost rejected the possibility of lack of economic incentives actually happening and found its potential impact on the pathway negligible. This was not the case with a perceived probable continuation of the economic recession and the consequent lack of financial capacity.

From a regulatory perspective, the main reservations regarded the *fuzziness/complexity of the regulatory and policy framework.* This is closely related to the instability of the political scene, but mostly reflects an ever-changing framework as a result of the deficit-amending modifications and the hitherto short-term nature of energy and climate action planning. What appeared to matter equally as much to stakeholders, in terms of severity and impact on the transition pathway, is the inherent *bureaucracy* of the processes necessary for incentive programme applications, energy upgrade certifications, etc. However, complicated bureaucratic processes appear to be more critical, given that all stakeholders deem that the capacity to mitigate their impacts is significantly better when compared to a fuzzy regulatory framework. This is why, with the exception of the representatives of the research community and of electric utilities, all stakeholder groups appeared to be significantly more concerned with regulatory bureaucracy than fuzziness.

On the technological axis, the main concerns expressed revolved around the current *lack of storage technologies*; poorly equipped, *inexperienced personnel*; and *geographic barriers*. The latter include difficulties in implementing actions in city centres, as well as poor infrastructure risks such as the saturation and security level of the power grid and the absence of interconnections necessary for proper penetration of renewables in the built environment.

Other implementation risks were only mentioned once or twice and as insignificant. These include the inadequacy of business models for the residential sector, plus the limited liberalisation of the internal electricity market and respective operators. An interesting example lies in the limited reference to the ageing Greek building stock, which could be considered either as a risk hindering the diffusion of energy-efficient technologies or overburdening the transformation costs, or as an opportunity for the success of the transition pathway (given the large underlying potential), depending on the perspective.

The identified implementation risks can also have synergistic effects. For example, poor prioritisation of climatic change and action, political instability, and frequent changes to a consequently fuzzy regulatory environment can all be intertwined, as well as linked to a sceptic, distrustful, even hostile society. Drawing from the unique characteristics of the Greek economy, another example can be found in the financial capacity. An ongoing recession can have detrimental impacts on all other dimensions and significantly expand the ground for the manifestation of the remaining implementation risks. This adds not only to the likelihood of their occurrence but also to the level of their impact and the capacity to mitigate either.

In the MCGDM analysis on the perceived performance of the ten implementation risks against their likelihood to occur, the level of their impact on the transition pathway, the capacity to mitigate their impacts, and the level of the concern, as perceived by the involved stakeholders, the latter appear to be mostly worried by political inertia, closely followed by the lack of financial capacity and the bureaucratic complexity of the energy efficiency-associated processes (Figure 11.1). They feel that both of the perceived risks on the societal axis are of medium importance and relevance to the effective design of a sustainable and robust pathway to an energy-efficient and climate-resilient Greek economy. Finally, non-provision for economic incentives, tax breaks, and subsidies, along with risks of technological nature, were considered the least critical by the engaged stakeholders.

Consequential risks

The *tariff deficit* is the most prominent concern in the discussion of potentially negative consequences of a low-carbon transition pathway, and of an energy

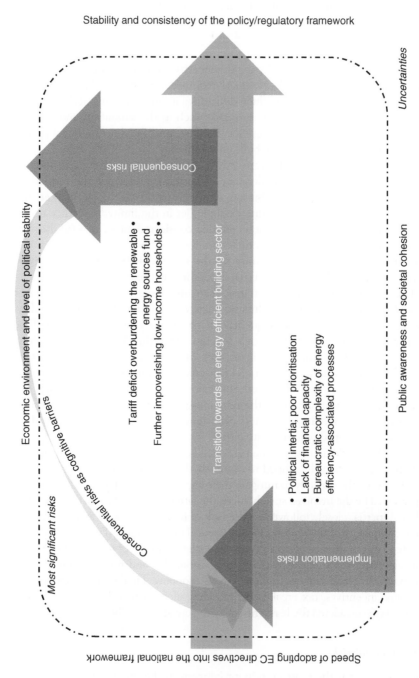

Figure 11.1 Main uncertainties, implementation, and consequential risks associated with an energy-efficient economy in Greece.

efficiency-oriented policy framework aimed at promoting this pathway. This focus can, to a large extent, be attributed to the sum of legislative mistakes made when designing the financial support mechanisms of the past. This refers to the pace of renewable energy penetration into the power generation mix, causing an equally high pressure of liquidity demands for compensating producers, based on the design of the FIT mechanism. Coupled with delays in relevant payments, these liquidity gaps created a large deficit in the Renewable Energy Sources Fund. Further legislative advances in an effort to reduce the deficit mostly revolved around tariff cuts and heavy taxation (Dusonchet and Telaretti, 2015), and eventually managed to freeze licensing applications and connection requests at the cost of solar power development and of public trust in the policy framework.

This case is an illustrative example of the life cycle of a barrier and the interplay between its implementation and consequential nature. Poor policy designs of the past, after giving rise to a large deficit in the renewables fund that in turn led to an ever-changing policy framework, eventually developed mistrust of the government for its ability to vote and retract policies and mechanisms. This is now largely considered as a significant implementation risk to any relevant action in the future. At the same time, the knowledge of such a consequential risk among policymakers and relevant stakeholder groups creates cognitive barriers to introducing mechanisms that can potentially allow such a risk to manifest. In other words, the perception and consideration of the consequential risk of developing a deficit in a support fund may also act as an implementation risk, hindering the design of appropriate support mechanisms.

Another risk that evidently concerns the stakeholders is potential implications for *poverty* as a by-product of climate and energy efficiency actions. Although an adverse economic environment is considered one of the main exogenous barriers to realising a low-carbon transition, stakeholders are largely worried about the possibility of such a transition further impoverishing Greek households. For example, according to an expert coming from the private-sector energy industry, well-funded yet poorly designed infrastructure and building renovation projects may lead to widening inequalities – the opposite result to that intended. Funding programmes and financial support must be strictly prioritised and planned in the long term in order for the transition to boost the economy, instead of aggravating the existing crisis. Furthermore, a stakeholder working in the banking sector acknowledged that, so far, financial incentives aimed at the residential sector primarily targeted lower-income households. These enabled them to apply for loans at a low interest rate, yet the loans became hard to pay in a time of recession. A similar approach may overburden the middle class and further impoverish the working class.

Finally, two consequential risks arose during the discussions on employment and investments. Some stakeholders mentioned the possibility of a low-carbon transition having negative implications for *employment*, especially with regard to the energy transformations brought about by further development of the solar power sector, e.g. in the fossil-fuel extraction and transformation sector. The interviewed policymakers practically rejected any notable impact, but nevertheless expressed their concern of such a probability. Although most experts agreed on the possibility that an energy-efficient transformation of the Greek economy might *discourage new investments*, about half of the engaged stakeholders considered that such an impact would be virtually non-existent. Besides, as most of the involved representatives of the banking and policymaking community noted, the capacity to mitigate such an effect is relatively high, provided that the policy strategies promoting the required transformations are effectively and sustainably designed.

Stakeholders compared these risks of the envisaged transition pathway along multiple criteria including the likelihood for them to occur and the severity of their impact. They also examined respective policy strategies with regard to these risks, including the capacity to mitigate their impact and the level of the stakeholders' concern. From their perspective, the most critical concern is the possibility of another tariff deficit, followed by a potential negative effect on poverty in the country. Following these, stakeholders collectively found that other negative socio-economic implications of actions towards an energy-efficient economy for private investments and unemployment are relatively insignificant.

Finally, stakeholders believe that sustainable future strategies, in a scenario where climate change remains contained, should be based on behavioural change. However, under scenarios with considerable mitigation and adaptation challenges, with higher probabilities of the identified implementation risks manifesting, they consider improving energy efficiency in the public building stock to be the most robust strategy (Nikas, Ntanos, and Doukas, 2019).

Conclusions

This narrative illustrates the uncertainties and risks associated with energy efficiency actions towards a future autonomous, nearly-zero-energy building sector, drawing from the expertise of and tacit knowledge embedded in experts coming from relevant stakeholder groups. Based on the conditions characterising an environment that is highly uncertain across the economic, regulatory. and socio-political axes, this research highlights the most important implementation and consequential risks of a transition pathway aiming to enhance energy efficiency. It does this over two different time horizons.

In the near term, stakeholders worry mostly about the current adverse market conditions, the bureaucratic nature of the processes associated with the implementation of measures and funding or incentivising investments, and the difficulties in aligning municipal priorities with the obligations of the central government. These are the main perceived implementation risks hindering Greek efforts to meet the national commitments on energy efficiency targets.

In the longer term, broader sets of risks appear to jeopardise the success of a low-carbon and energy-efficient transition of the Greek economy. On one level, a multitude of implementation risks are identified, the most important of which appear to concern poor prioritisation and political inertia as a result of an ongoing political crisis and, once again, a bureaucratic and complex regulatory framework coupled with limited financial capacity to incentivise and invest in energy efficiency actions. On another level, stakeholders' concerns over the potentially negative consequences of such a pathway appear to almost exclusively focus on the recurrence of a tariff deficit and the possibility of further impoverishing lower-income households.

It should be noted that the study is based on two different series of stakeholder engagement approaches, involving different stakeholders. This differentiation may be another reason, apart from different considerations of the time horizon, why so many differences were observed between the sets of risks identified. In fact, discussions on near-term energy efficiency targets were carried out only with stakeholders from the policymaking group; bringing in representatives from other expert groups and exploring differences in perceptions across all stakeholder groups would add valuable insights into the discussion.

The participation of a diverse set of stakeholders in the long-term perspective part of the study allowed for such differences to be highlighted, providing a broader picture of risk perception on the Greek energy efficiency front. However, only three of the original policymaking group participated in the second part of the study, and this may have been instrumental in obtaining such a differentiation in risk mapping and evaluation. Another reason behind this may be found in the abstractness of the policy framework discussed on the long-term perspective. When they were aware of the exact policy instruments, stakeholders appeared capable of matching each measure with its associated risk and were therefore provided a better context to identify the risks. Discussions on the long-term vision, by contrast, oriented on the transition pathway rather than the policies promoting it. The study also lacked the iterative dimension that would enable consensus building among the participating stakeholders. Finally, although the discourse mainly draws from detailed discussions with the engaged stakeholders, the multi-criteria analyses rely on questionnaire-like evaluations, for which a significantly larger sample size could have been used.

In order to reach its full potential, this kind of research can be linked to quantitative risk assessment. The more standardised results underlying this narrative have been used as a basis for other analyses, namely a portfolio analysis for optimising the near-term policy mix and a fuzzy cognitive mapping study for exploring different long-term transition strategies. Further potential lies in their transformation and incorporation into quantitative systems' model-driven analyses, to produce fruitful insights with specific policy implications.

References

Capros, P., De Vita, A., Tasios, N., Siskos, P., Kannavou, M., Petropoulos, A., Evangelopoulou, S., Zampara, M., Papadopoulos, D., Nakos, C., Paroussos, L. (2016). *EU Reference Scenario 2016: Energy, transport and GHG emissions Trends to 2050*. Publications Office of the European Union, Luxembourg.

Centre for Renewable Energy Sources and Saving (2015). *Energy efficiency trends and policies in Greece*. Athens.

Doukas, H., Nikas, A. (2019). Decision Support Models in Climate Policy. *European Journal of Operational Research*, in press.

Doukas, H., Karakosta, C. and Psarras, J. (2009). A linguistic TOPSIS model to evaluate the sustainability of renewable energy options. *International Journal of Global Energy Issues* 32(1–2), 102–118.

Doukas, H., Karakosta, C., Flamos, A., Psarras, J. (2014). Foresight for energy policy: techniques and methods employed in Greece. *Energy Sources, Part B: Economics, Planning, and Policy* 9(2), 109–119.

Dusonchet, L., Telaretti, E. (2015). Comparative economic analysis of support policies for solar PV in the most representative EU countries. *Renewable and Sustainable Energy Reviews* 42, 986–998.

European Commission (2008). *Europeans' Attitudes towards Climate Change, Special Eurobarometer 300/Wave 69.2*. Brussels.

European Commission (2013). *Climate Change, Special Eurobarometer 409/Wave EB80.2*. Brussels.

Fawcett, T., Rosenow, J., Bertoldi, P. (2017). *The future of energy efficiency obligation schemes in the EU*. ECEEE 2017 Summer Study – Consumption, Efficiency & Limits.

Forouli, A., Gkonis, N., Nikas, A., Siskos, E., Doukas, H., Tourkolias, C. (2019). Energy efficiency promotion in Greece in light of risk: Evaluating policies as portfolio assets. *Energy Policy*, 170, 818–831.

Gaglia, A.G., Tsikaloudaki, A.G., Laskos, C.M., Dialynas, E.N., Argiriou, A.A. (2017). The impact of the energy performance regulations updated on the construction technology, economics and energy aspects of new residential buildings: The case of Greece. *Energy and Buildings* 155, 225–237.

HELAPCO (2016). *Statistics of the Greek photovoltaic market in 2015*. Hellenic Association of Photovoltaic Companies, Athens. Available at: http://helapco.gr/wp-content/uploads/pv-stats_greece_2015_1Mar2016.pdf.

Hellenic Statistical Authority (2011). *Buildings' construction per time period*. Available at: www.statistics.gr/documents/20181/1204362/A1601_SKT01_TB_DC_00_2011_05A_F_GR.xls/b14a676e-3d69-49c3-ba5e-11665b23e536.

Kaplanoglou, G., Rapanos, V.T. (2018). Evolutions in consumption inequality and poverty in Greece: The impact of the crisis and austerity policies. *Review of Income and Wealth* 64(1), 105–126.

Ministry of Environment and Energy (2014). *3rd National Energy Efficiency Action Plan of Greece*. Athens.

Ministry of Environment and Energy (2017). *4th National Energy Efficiency Action Plan of Greece*. Athens.

Nikas, A., Doukas, H. (2016). Developing robust climate policies: a fuzzy cognitive map approach, in: *Robustness Analysis in Decision Aiding, Optimization, and Analytics* (239–263). Springer, Cham, Switzerland.

Nikas, A., Doukas, H., López, L.M. (2018). A group decision making tool for assessing climate policy risks against multiple criteria. *Heliyon* 4(3), e00588.

Nikas, A., Ntanos, E., Doukas, H. (2019). A semi-quantitative modelling application for assessing energy efficiency strategies. *Applied Soft Computing*, 76, 140–155.

Nikas, A., Doukas, H., Lieu, J., Alvarez-Tinoco, R., Charisopoulos, V., van der Gaast, W. (2017). Managing stakeholder knowledge for the evaluation of innovation systems in the face of climate change. *Journal of Knowledge Management* 21(5), 1013–1034.

Papoulis, D., Kaika, D., Bampatsou, C., Zervas, E. (2015). Public Perception of Climate Change in a Period of Economic Crisis. *Climate* 3(3), 715–726.

Simoglou, C.K., Bakirtzis, E.A., Biskas, P.N., Bakirtzis, A.G. (2018). Probabilistic evaluation of the long-term power system resource adequacy: The Greek case. *Energy Policy* 117, 295–306.

Spyridaki, N.A., Ioannou, A., Flamos, A., Oikonomou, V. (2016). An ex-post assessment of the regulation on the energy performance of buildings in Greece and the Netherlands: a cross-country comparison. *Energy Efficiency* 9(2), 261–279.

Stavrakakis, Y., Katsambekis, G. (2018). The populism/anti-populism frontier and its mediation in crisis-ridden Greece: from discursive divide to emerging cleavage? *European Political Science*, 1–16.

Tselepis, S. (2015). The PV Market Developments in Greece. *Net-Metering Study Cases*. 28th EUPVSEC, Paris.

Tsita, K.G., Pilavachi, P.A. (2017). Decarbonizing the Greek road transport sector using alternative technologies and fuels. *Thermal Science and Engineering Progress* 1, 15–24.

Zerva, A., Tsantopoulos, G., Grigoroudis, E., Arabatzis, G. (2018). Perceived citizens' satisfaction with climate change stakeholders using a multicriteria decision analysis approach. *Environmental Science & Policy*, 82, 60–70.

Part V

Pathways focusing on renewable energy technologies at household and community levels

12 Indonesia

Risks and uncertainties associated with biogas for cooking and electricity

Mariana Silaen, Yudiandra Yuwono, Richard Taylor, Tahia Devisscher, Syamsidar Thamrin, Cynthia Ismail, and Takeshi Takama

Introduction

Indonesia's energy needs and policy ambitions

Indonesia's significant economic growth in the past decade, due to increasing investment and export growth, has been accompanied by increasing energy demand that is met by oil, gas, coal, and renewables (World Bank, 2018; Secretariat General of National Energy Council, 2016: 51). Of these, coal is Indonesia's most important energy resource for domestic power production and also garners vast export earnings. Nearly 80% of the nation's coal productions were exported in 2015, making it the largest coal exporter in the world despite possessing only 2.2% of the global coal reserves (Atteridge, Aung, and Nugroho, 2018; BP, 2018; Cornot-Gandolphe, 2017). As a result, the increasing energy demand and consumption have made Indonesia the eighth-largest greenhouse gas (GHG) global emitter (Friedrich, Ge, and Damassa, 2015). In 2012, CO_2 emissions were distributed across three main sectors, with power generation, industry, and transportation accounting for roughly 33%, 30%, and 29% respectively (Republic of Indonesia, 2015).

In 2016, Indonesia introduced its Nationally Determined Contribution (NDC) targets of a 26% and 29% GHG emission reduction by 2020 and 2030 respectively, in comparison to the business-as-usual scenario. However, capacity additions have been in favour of coal and most of the planned renewables come in well after 2020 (Climate Action Tracker, 2018). A high-carbon pathway, evidenced in these energy plans and GHG emission trends, is far from being consistent with NDC targets. Indonesia's policies are currently rated as 'highly insufficient' to meet its NDC targets (Climate Action Tracker, 2018). Like other countries, it faces several challenges to mainstream and integrate climate change into national planning and development processes.

Indonesia's main strategy for development is formulated in the National Long-Term Development Plan (RPJPN) which is divided into four five-year National Medium-Term Development Plans (RPJMN). The current RPJMN applies from 2015 up to 2019. One of its aims is to increase the contribution of renewable energy to help move the renewable energy share to 23% of total primary energy supply by 2025 as mentioned in the NDC. Although the target for the national energy exists, there is no clear implementation plan from the government for how the country will meet the goal. Moreover, other policies in the energy sector often run counter to these commitments, suggesting that mainstreaming is challenged by other priorities. For instance, Indonesia has started shifting coal from international markets to meeting domestic energy demand (IEA, 2014). These targets need concerted efforts and strong support from the government to better integrate emissions and renewable energy plans into energy policy frameworks.

Indeed, policy support for fossil fuel undermines meeting both the NDC targets and the SDGs as the fossil-fuel sector plays a prominent role in economic development for Indonesia. Shifting to a more sustainable economic pathway including clean energy production faces many barriers as economic growth is prioritised over other issues. This contradiction is portrayed by the ongoing construction of non-renewable power plants. Over the past five years, coal capacity has increased by around 13.6 GW compared to only 1.8 GW of renewable energy (Climate Action Tracker, 2018). Furthermore, the new power plants do not use the most modern energy-efficient technologies. The expansion of coal mining also risks the potential lock-in of carbon-intensive infrastructure and financial assets (Atteridge, Aung, and Nugroho, 2018). As one of the world's largest coal exporters, Indonesia stands to lose much of this revenue when other countries implement their own mitigation measures. Other problems, such as low tax revenue and low commodity prices, combined with complex bureaucratic and transparency issues, also hinder clean energy infrastructure investment (OECD, 2015).

Rethinking renewable energy solutions in Indonesia

Renewable energy solutions, including bioenergy, need to be geographically and culturally appropriate, low-cost clean fuels that meet energy needs and provide co-benefits; they also need to be practically advantageous by offering, for example, simple implementation, good technological availability, and a strong base of experiential knowledge (Blenkinsopp, Coles, and Kirwan, 2013; Brent and Kruger, 2009; Urmee and Md, 2016). Expanding the use of renewable energy sources (RES) is essential to meet future domestic energy demands and to achieve policy targets. In Indonesia, hydro, geothermal, and biomass are among the promising sources to develop further. However, to date the combined installed capacities of these alternatives remains approximately 7.8% of the total optimum capacities (cited in IEA, 2015).

Bioenergy, the initial object of interest in this narrative, offers many options that have not yet been widely used. Furthermore, the greater availability of land, favourable climatic conditions for agriculture, and lower labour costs also support this focus (Widodo and Rahmarestia, 2008). Focusing on biomass, the nation's bioenergy comes in many forms of value chains, such as biomass gasification (Asadullah, 2014), which have been developed so far with the power capacity of 10–100 kWe (Singh and Setiawan, 2013). Additional potential value chains include synthetic gas using pellets from various species of trees and palm-oil solid waste (Kusumaningrum and Munawar, 2014; Siregar *et al.*, 2017) and anaerobic digestion using agricultural and livestock waste. The latter has been considered successfully implemented by a programme called BIRU (see BIRU, 2018).

Biogas as the focus of this chapter, especially when produced through agricultural waste, is one viable alternative since it can be implemented in rural, and sometimes remote, areas where many of Indonesia's populations reside and make a living in small-scale agriculture. Biogas provides GHG emission mitigation benefits by lessening demand usage of conventional energy. According to an estimation by BIRU, two million small biogas digesters could potentially be installed in Indonesia, being equivalent to a reduction of 6.4 million tons CO_2/year (cited in Devisscher *et al.*, 2017). Meanwhile, the estimated potential capacity for large-scale biogas-to-electricity production is 2.6 GW (Government of Indonesia, 2017). For this reason, biogas pathways are of increasing interest to policymakers because of their carbon and energy benefits and numerous potential co-benefits, such as suppressing unmanaged firewood collection, promoting waste management, and helping with the use of biogas slurry as organic fertiliser (Bedi, Sparrow, & Tasciotti, 2017).

Overall, biogas offers some promising practical and feasible alternative energy options for Indonesia. But how could this technology be rolled out across as ethnically and culturally diverse country as Indonesia, spread across more than 17,000 islands? What about the suitability, the social acceptance, and the gender and equity dimensions of biogas? What do stakeholders see as the main opportunities and risks of biogas, and what can the latest research tell us? This narrative investigates the potential of biogas to help meet domestic energy needs and to comply with Indonesia's climate mitigation commitments and development planning. At the core are experiences with four biogas programmes taking place in Bali; each having different motivations, practices, outcomes, and lessons. A better comprehension of the risks and uncertainties associated with biogas development pathways can help support future dialogue and planning on climate, energy, and development.

Research process and methods

A range of stakeholder engagement activities were used to understand how stakeholders perceive and manage risks. Risks were discussed during different types of meetings, using different research methods as described in Table 12.1.

Activities included a series of workshops and a period of in-depth fieldwork. The team also conducted continuous dialogues with national-level policymakers to discuss the pathways of choice. The dialogue was a result of a collaboration with the Ministry of Development and Planning (Bappenas), which also involved the Ministry of Energy and Mineral Resources (ESDM) and Indonesia's state-owned electricity company (PLN), which runs the nation's monopoly market in electricity distribution and generates the majority of the country's electrical power. We made use of the bioenergy workshops to present and gain information about the pathways and conduct research validation. We also involved additional stakeholders, including an NGO called Yayasan Rumah Energi (YRE), subnational agricultural and energy agency officers, local farmers (users and nonusers), researchers from the local university, banks, and business/private sectors.

We initially focused on the small-scale biogas for rural households as the first priority chosen by the participants. Further potential pathways were formulated based on the earlier stage of the research, as well as the economic and political drivers in Indonesia (Devisscher *et al.*, 2017). Later on, interest divided to a large-scale biogas-for-electricity production pathway and this was discussed during the third workshop and policy dialogues. More information about the research methods used, the development of a 'toolkit' for data collection and analysis, and the applicable framework for integration of concepts and methods can be found in Devisscher *et al.* (2017).

Table 12.1 Chronology of events and activities and research methods

Event or activity	Stakeholders involved	Methods applied
1st Bioenergy Workshop on 'Scoping and Envisioning', Bali, May 2016	Local farmers, NGOs, local and national government officials, research agencies, banks	Multicriteria assessment for priority technology identification for bioenergy, H-form, system mapping
Field work in Bali and Jakarta, October–November 2016	Local farmers, NGOs, local government officials	Focus group discussions, interviews
2nd Bioenergy Workshop on 'Solutions, Business Models and Enabling Conditions', Bali, May 2017	Local farmers, NGOs, private sectors, local government officials, research agencies	Focus group discussions and exercises for Q-methodology
Policy dialogues, Jakarta, February–March 2018	Government officials from Bappenas, ESDM, and PLN	Focus group discussions, risk identification and assessment exercise, pathway exercise
3rd Bioenergy Workshop on 'Green Business and Synergy Action', Bali, April 2018	Local farmers, NGOs, local and national government officials, research agencies	Focus group discussions, exercises on two biogas pathways and Q-methodology validation

Transition pathways

Focused on finding ways to meet energy needs, NDCs, and development targets, the Indonesia TRANSrisk narrative aims at improving the understanding on how biogas alternatives could effectively contribute to a low-carbon energy transition and what changes are required to achieve it. Bali is the target area due to its high potential feedstock for bioenergy production, which could support up to 30% of its current power plant capacity (Government of Indonesia, 2017; Kementerian ESDM, 2017). Furthermore, several biogas programmes have operated in Bali, especially at the small-scale level. These programmes were implemented by the Bali Provincial Agricultural Agency (i.e. SIMANTRI programme), the Agency of Public Works, the West Bali National Park, and an NGO called Yayasan Rumah Energi (YRE, i.e. the BIRU programme). All programmes installed individual biogas digesters except SIMANTRI, which carried out communal installations. Also, the government's programmes went for fully subsidised operations while the BIRU programme used a market-based approach with partial subsidies (Devisscher *et al.*, 2017). The guarantee period of SIMANTRI and Public Works programmes was set to three months, while BIRU's was up to three years including maintenance services. However, no clear guarantee scheme existed in the West Bali National Park pilot project. Furthermore, the SIMANTRI programme targeted economic development for the farmers' livelihood, in which biogas was a supporting system of the whole project; and the Public Work's programme aimed to support the national policy mandate of renewable energy deployment. Both the others aimed at addressing environmental issues including carbon emissions, with the BIRU programme also focused on energy access and West Bali National Park on reducing forest degradation.

The two selected pathways in this chapter explore options for a low-investment/short-term scenario and a high-investment/long-term scenario. The first is an easily implementable, low-cost household-scale biogas digester system supplying household energy needs. This pathway also foresees the transfer of these systems and the know-how to other geographical areas. The second pathway focuses on large-scale biogas systems that produce electricity, require higher investment, and generate high benefits in the long run. The first pathway uses experiences by farmers (biogas adopters) in Bali, while the second one was developed in discussions with policymaker experts from Bappenas (Ministry of National Development Planning), PLN (Electricity Company), and ESDM (Ministry of Energy and Mineral Resources).

Household biogas transition pathway

The household biogas transition pathway is concerned with meeting domestic energy needs for cooking and lighting, which also intersects with issues of health in rural areas, community social structure, as well as smallholder productivity. These issues are important for understanding the pathway.

Nearly one-third of Indonesia's working population consists of farmers in rural areas (BPS, 2017) where solid fuels are mostly used for cooking and are often associated with health problems (Gall *et al.*, 2013). Indoor pollution in the home from solid fuels utilisation contributes to respiratory infections and diseases. Biogas for household cooking and lighting is clean and safe while also fitting the profile of the rural areas. It can also provide fuel for cooking, give the benefit of waste management, reduce the indoor pollution risk, produce biogas slurry for organic fertiliser, and even provide lighting where possible.

With low population density in Indonesian rural areas, a household-level or individual system to provide fuel for cooking and lighting could potentially be a better option than a centralised or communal system that requires greater material input as well as management co-ordination. Moreover, an individual system could be deployed without having to formulate a gas (or electricity) distribution system plan within the community, thus potentially reducing the amount of investment.

In addition, biogas offers a potential means to increase farmers' resilience by, for instance, the use of biogas slurry as organic fertiliser. These could generate new sources of supplementary or additional income for the smallholder, who may trade organic coffee beans with organic fertiliser made on the premises. These benefits to the smallholder farmers should be added to the savings made from reducing the reliance on fossil-fuel-based energy. Additionally, if the government decided to suppress the liquid petroleum gas (LPG) subsidy, those savings will be further increased. These potential benefits add value to the pathway, not only to help the country achieve its NDC target by reducing rural households' fossil-fuel consumption but also to increase people's resilience.

Large-scale biogas-for-electricity transition pathway

The large-scale biogas transition pathway is mainly concerned with meeting renewable energy and carbon emission targets, as well as a means of addressing the development goal of providing electricity access in remote areas. It has economic potential for the creation of small energy-generating enterprises.

Since 2014, the government of Indonesia has focused on increasing electricity access to rural areas, including remote islands. As it stands, Indonesia has achieved above 94% of its electrification ratio target of 92.75% in 2017 through the Energy-saving Solar-powered Lighting Supply (LTSHE) programme in rural areas (Kementerian ESDM, 2017), although doubts remain over the sustainability of the programme. However, and in contrast, the majority of Indonesia's current and planned power plants are coal-powered generators. A new approach to investment in electricity generation that has longer-term objectives of supporting low-carbon development may be needed. Currently, one of the most notable initiatives comes from PLN which, under ESDM policy, is opening its doors to private companies by purchasing their services to operate in different areas to generate and sell electricity.

Responding to this policy, many private sector actors are working on renewable energy initiatives but they are less active in biogas-to-electricity enterprises. Stakeholder dialogues elaborated that the current policy is not totally supportive of electricity generation from the RES due to their higher initial and operating costs, especially from biomass. Thus, companies selling electricity generated by biogas will face challenges when competing with the existing electricity price set by the government. There was, however, growing political support to improve the policy, such as by planning for electricity production using renewable energy and strengthening the feed-in tariff implementation.

Currently, biogas plants are utilised in some areas of Indonesia that are claimed to have performed successfully both as waste management reactors and energy generators. Based on these experiences, the government representatives acknowledged that an approach to increase the number of larger communal biogas plants to produce electricity would be beneficial. They indicated that large biogas plants could potentially support electricity accessibility, apart from its main goal of reducing the emissions from the power generation sector.

Risks and uncertainties associated with biogas development in Indonesia

It is important to first note the subjective nature of risk perception. How risk is experienced and understood encompasses personal arguments of possible events and their consequences (Aven, 2012). What may pose a risk for one group of stakeholders may be totally satisfactory and unproblematic to another. For example, government researchers identified unfiltered H_2S (hydrogen sulphide) in biogas installations as a potential danger, while farmers did not recognise it as a risk due to their different concerns and the lack of knowledge about such facts.

To mitigate some subjectivity, this narrative utilises a widely understood and shared category-based risk assessment to summarise and convey risks. Fourteen risks were identified and formulated into six risk categories (Table 12.2). From this, it is clear that the highest number of risks are associated with technological

Table 12.2 Classified risk perceptions

Social	Technological	Economic
1 Collective management issues in the case of larger biogas systems 2 Imbalance time spent between men and women	3 Accessibility and sustainability of feedstock 4 Poor maintenance and infrastructure 5 Unfitting technology choice for local conditions 6 Leakage of methane gas emissions	7 Lack of long-term incentives to aim for value-added activities 8 Subsidy on fossil fuels

Environmental	Political	Regulatory
9 Unfiltered hydrogen sulphide in biogas installations	10 Different goals on biogas installations 11 Varied monitoring practices in different programmes 12 Initial investment needs	13 Lengthy and bureaucratic process to apply for support form biogas programmes

aspects such as feedstock, infrastructure, and electricity reliability. These risks will later be introduced as implementation risks, also understood as barriers, and consequential risks, also understood as negative outcomes (Figure 12.1). All policy decisions and actions carry some potential risks that need to be considered in the context of multiple uncertainties.

Uncertainties in the development of biogas as a renewable energy option

Four key uncertainties underlying the energy transition to biogas were identified: (i) the unclear role of public and private sectors; (ii) the changing focus over each political term; (iii) the unspecified national biogas target; and (iv) different views on biogas development. Otherwise understood as 'epistemic uncertainty', the first three risks are a result of incomplete or insufficient knowledge of the system, while the fourth one is rather classified as an 'ambiguity' due to the existence of multiple knowledge frames.

First, the blurred lines between the roles of public and private sectors are responsible for a situation where overlapping and even contrasting context and goals of biogas development exist. This undermines the ability of all actors involved to form innovative partnerships, intensify engagement, or effectively collaborate on programmes. Second, the end of each political term in Indonesia creates a significant uncertainty as there is an absence of knowledge whether the new government will continue or abandon previous/existing programmes. Both of these uncertainties are compounded by a third uncertainty: the lack of biogas development targets. This lack of leadership from the government affects commitments to biogas development by the civil service and the private sectors. If the national biogas target was set, it would motivate ministries and other actors to better co-ordinate biogas programmes to pursue a common objective (Devisscher *et al.*, 2017). Fourth, we found three different perspectives or 'worldviews' on the development and value of biogas. They include a technological-based paradigm in addressing barriers, a scale-based paradigm that put forward the agricultural benefits, and independence prioritisation through biogas. The lack of consensus and differences in viewpoints, framings, and perspectives may be implicit or tacit, thus adding additional layers of uncertainty. This uncertainty is present in the different objectives of biogas programmes in Bali, for example the BIRU programme that undertook market-based approach that emphasised technological services and the SIMANTRI programme that put forward the sustainable agriculture and economic profit approach.

Out of the four sources of uncertainty, three stem from the influence of government. The fourth source of uncertainty is a consequence of institutional complexity and the diversity of actors and their experiences and priorities. Investigating this complexity can nonetheless reveal important patterns, such as distinctive views on biogas development and the ways in which different constellations of actors line up behind these perspectives.

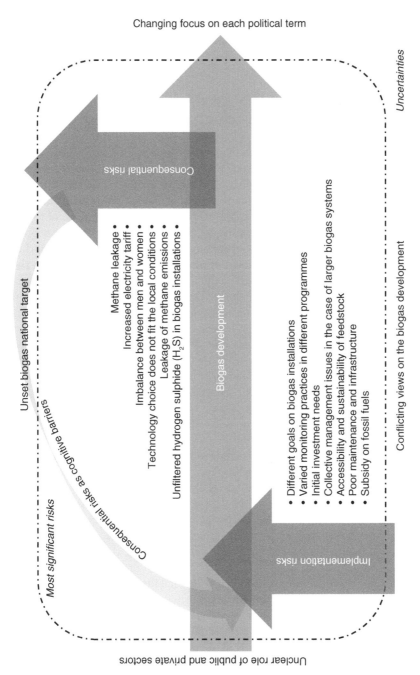

Changing focus on each political term

Consequential risks

- Methane leakage
- Increased electricity tariff
- Imbalance between men and women
- Technology choice does not fit the local conditions
- Leakage of methane emissions
- Unfiltered hydrogen sulphide (H_2S) in biogas installations

Biogas development

Implementation risks

- Different goals on biogas installations
- Varied monitoring practices in different programmes
- Initial investment needs
- Collective management issues in the case of larger biogas systems
- Accessibility and sustainability of feedstock
- Poor maintenance and infrastructure
- Subsidy on fossil fuels

Unset biogas national target

Consequential risks as cognitive barriers

Most significant risks

Unclear role of public and private sectors

Conflicting views on the biogas development

Uncertainties

Figure 12.1 Main uncertainties, implementation, and consequential risks associated with biogas development in Indonesia.

These findings illustrate the possible barriers that policy actors and development planners need to be aware of and sensitive to when designing biogas programmes. To be acceptable to stakeholders with a plurality of perspectives, plans need to include different ways of providing benefits and incentives that motivate participants, while mitigating the perceived drawbacks.

Implementation risks of biogas development in Indonesia

Different motivations as a result of different 'worldviews' of biogas development were fundamental causes of further incoherence in the monitoring processes, there being no overall consensus on biogas monitoring procedures. We revealed that most of the rural biogas digesters had been abandoned due to the lack of investment in monitoring and evaluation (Devisscher *et al.*, 2017). Monitoring practices are required to assess the current situation of the programmes in order to ensure their sustainability and to help identify technical issues, one of which includes proper infrastructure maintenance. However, appropriate maintenance is lacking in the government-led programmes, which is identified as another risk. When farmers faced constraints in operating biogas or suffered faults in the technology, they gave up using it, especially in the cases where they had no warranty and were not technically trained (Devisscher *et al.*, 2017). Moreover, highly subsidised digesters with little investment in maintenance and monitoring were less successful due to a lower sense of ownership of the biogas digester and a trade-off with the maintenance cost. The representative of ESDM stated that the adequacy and reliability of the overall technological infrastructure should be able to reduce the risks where maintenance is lacking.

In addition, the government-run programmes also presented a barrier in terms of the biogas technology distribution to beneficiaries. The farmers as the targeted beneficiaries stated that the process of obtaining a biogas digester was overly bureaucratic and time consuming (Devisscher *et al.*, 2017). In other words, there are various stages farmers need to undergo to obtain the digesters and, even when offered with no costs, this acts to prevent the farmers from adopting and fully utilising the technology.

Furthermore, accessibility to feedstock was also considered to be a barrier by both users and policymakers (Devisscher *et al.*, 2017). This risk is worth being discussed because, as repeatedly emphasised, the abundance of feedstock is often given as a main justification for biogas development. Yet the farmers reported feedstock shortages, besides considering the collection process as time consuming. The barrier concerning the feedstock availability is also noticed by the ESDM representatives with a similar reason; thus they recommended a system to ensure the feedstock sustainability.

Some risks specifically relate to communal biogas installations. In the SIMANTRI programme, barriers arising include management issues of the digesters (Devisscher *et al.*, 2017). Unsuitable and ineffective management will create risks in biogas deployment since in many cases the farmers did not seem to be committed to managing the biogas as a team. This appears to be correlated with the lack of familiarity with and sense of belonging to a relatively new technology. Another contributing factor to limited use of communal biogas systems was the location of digesters, which is often far from the communities. A consequence is that they only use biogas for shared needs such as water heating for drinking during community meetings, but not to benefit from cooking in their households.

Finally, a further barrier was suggested to stem from the lack of long-term incentives to engage in household biogas. Some stakeholders, especially policymakers, perceived the need for biogas adoption to go beyond 'cooking only' and rather to target value-adding activities to keep farmers engaged in the long run. This sentiment is related to the 'outscaling' pathway suggested in terms of co-benefits generated that can become increasingly attractive in the longer term; the policymakers questioned the potential of household biogas to deliver these benefits.

The high initial investment cost is a further barrier or implementation risk, according to the representative of Bappenas. Whether sufficient government funding is available to meet the investment costs for renewable energy is dependent on the political situation. However, the future situation and its impact on investment are far from predictable. This type of investment relies on clear renewable energy targets, an appropriate policy framework, and strong and clear co-operation between public and private actors, where each actor plays its own key roles (Masini and Menichetti, 2012). In this case, the role of policymakers is to create incentives in order to achieve effective investment (IEA, 2007), while the private sector is expected to buffer the financial requirement towards the low-carbon economy (Masini and Menichetti, 2012). However, while the roles of the public and private sectors are either overlapping or leaving gaps, the investment risk remains higher. The risk on the initial investment is thus closely related to the three identified uncertainties that stem from the government influence.

The representatives of ESDM, Bappenas, and PLN stated their specific interest in larger-scale biogas and electricity generation. However, the government and the electricity companies have made limited advances in generating electricity from biogas. Concerns over technology development and infrastructure included the limited capacity of biogas-to-electricity plants to generate energy and electricity tariffs. The representative of PLN was concerned about whether or not electricity generation would be as sustainable in terms of continuity of feedstock supply. This needs to be ensured before buying the electricity from the providers, especially when entering into a long-term contract or partnership. Feed-in tariff schemes could play a critical role in this regard. Also, a PLN representative suggested that if the regulations were weak and technology remained inadequately developed, then production costs would continue to be too high for wide market penetration. The high associated production cost may also result in a higher electricity tariff and thereby be passed on to the end users. Under this scenario, electrification in remote areas would become very difficult to achieve.

Fossil-fuel subsidies present another challenge because they work against biogas development. Challenges posed are fossil fuels being more affordable compared to renewable energy systems. As stated by the Bappenas representatives, farmers, local businesses, and policymakers alike are more likely to favour fossil fuels. Nevertheless, in many rural or isolated areas subsidised fuels are not accessible or rarely available, causing the price to constantly fluctuate.

Consequential risks of biogas development in Indonesia

Environmental aspects were a central concern among the consequential risks. For example, some rural biogas digesters were not installed with H_2S filtering, which may harm the environment or even human and livestock health according to researchers in Udayana University (Devisscher *et al.*, 2017). It also runs the risk of corrosion to the digesters (Chaiprapat *et al.*, 2011). Large quantities of manure in the biodigester may release this gas, unless it is fitted with filtering technology. Unfortunately, in many cases the observed biogas digesters were not equipped with this technology.

Similarly, some methane leakages in the biogas digesters might occur when users do not burn the biogas produced. There have been some debates on whether or not the methane leakage is less harmful to the atmosphere than abandoning the manure on the ground or in the barns. Bruun *et al.* (2014) argued that a small amount of methane leakage could offset the emissions savings from faulty or improperly used digesters. Also, from the perspective of technological efficiency, biogas technology should be used as optimally as possible in order to take advantage of the feedstock and to reduce the overall GHG emission.

Besides the environmental risks, we identified a specific consequential risk within the social domain. The women in rural Bali tend to have significant roles in collecting the firewood and providing meals, while men's roles are mainly taking care of livestock and managing organic waste as feedstock for biogas technology. Substitution of biogas for firewood therefore has the effect of reducing the women's working time while increasing the men's. Such a role reversal has a positive aspect because women and their families can benefit from women spending more time doing other things. However, it was also recognised that this carries a risk of imbalance and turmoil in the household. Some debates arose around this perception as it is mostly a viewpoint of male farmers, while other stakeholders, such as the policymakers and some researchers, do not view it as a significant issue, suggesting it does not represent the entire situation in Indonesia. Yet such gender-based role division is embedded in many customs across Indonesia.

Despite interesting discussions on the gender role, the representative of PLN seemed to be more concerned about the other pathway, especially the reliability of electricity generation. Conditions may induce a risk to system stability within the interconnection grid. The current state of biogas production tends to fluctuate, which may disturb the current system. To resolve this, the PLN representative

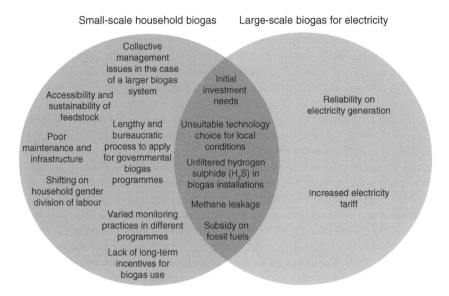

Small-scale household biogas   Large-scale biogas for electricity

Collective management issues in the case of a larger biogas system

Accessibility and sustainability of feedstock

Poor maintenance and infrastructure

Shifting on household gender division of labour

Lengthy and bureaucratic process to apply for governmental biogas programmes

Varied monitoring practices in different programmes

Lack of long-term incentives for biogas use

Initial investment needs

Unsuitable technology choice for local conditions

Unfiltered hydrogen sulphide (H_2S) in biogas installations

Methane leakage

Subsidy on fossil fuels

Reliability on electricity generation

Increased electricity tariff

Figure 12.2 Risks on both pathways: household biogas and large-scale biogas for electricity.

recommended three potential solutions: (i) utilisation of a smart grid; (ii) a co-generation/hybrid system; and (iii) the biogas system for off-grid or isolated areas.

Comparison of pathways

Both pathways aim to assist the government of Indonesia in meeting its emission reduction targets while providing energy access. Against these targets, this section will apply a cross-analysis by classifying each risk under each pathway. The interest in risks on the biogas-to-electricity pathway were stated by the policymakers after the detailed investigation on the household biogas had been delivered to them. Therefore, the risks on large-scale biogas were not identified in depth. Nevertheless, we obtained at least higher-level perceptions on this pathway. The risks from both pathways were either identified as exclusive or overlapping to each other (Figure 12.2).

Risks associated with both pathways

Chief among the risks associated with both pathways is the high initial investment needs (Figure 12.2). This was particularly on the minds of policymaker stakeholders. A representative of Bappenas suggested that the investment cost is a risk to both the small-scale household pathway or the large-scale electrification pathway as limited financial or monetary reliability (with the possibility of a budget constraint being subsequently imposed) will create obstacles for the overarching development. The lack of investment is likely to create a domino effect on supporting measures such as installation, maintenance, training, and incentives. The risk is identified in both pathways and heavily depends on the national policy on renewable energy. Initial investment is also shaped by enabling policies and regulations concerning the private sector. For example, the latest electricity law (UU no. 30/2009) and its implementing regulation (PP No. 14/2012) allowed the private sectors to be actively involved in the power sector, such as by running a power generation business (Kuvarakul *et al.*, 2014). Suitable laws and regulations would mitigate this risk by attracting more private-sector investment in either pathway, thereby financially supporting the low-carbon economy, as suggested previously (Masini and Menichetti, 2012).

Other important risks are related to how the comparative advantages of the main alternative energy options stack up. In this respect, one of the main challenges is the cheaper fossil-fuel energy. For example, a three-kilogram LPG tank for households is subsidised, as regulated in ESDM's ministerial decree no. 2458 K/12/MEM/2017. The effect of such subsidy should not be underestimated as, for many rural residents that use LPG, there is a risk of preferring LPG to biogas in this pathway. The main causes are cheaper conventional energy with lower initial investment at the household level, as well as fewer technological barriers. When it comes electricity generation, PLN also faces challenges in purchasing the renewables in general since they are more costly than coal, which is subsidised by the government (International Institute for Sustainable Development, 2018). This subsidy itself is one of the rooted reasonings behind the more affordable current electricity tariff.

Other risks relevant to both pathways concern the choice of technology, which is not always appropriate. For instance, it was noted that in some cases household biogas installations had developed cracks or leaks in the tanks due to unsuitable local biophysical conditions (Devisscher *et al.*, 2017). Taking into account the local conditions is a necessity according to the policymakers who emphasised that not all Indonesian regions are suitable for biogas. Bappenas gave the example of Nusa Tenggara Timur (NTT) province – a semi-arid area where biogas development would be difficult, since continuous water supply is essential to operate the technology. On the other hand, the island of Sumatra, one of the world's largest palm-oil producers, would make a promising biogas generator (Rahayu *et al.*, 2015). The abundance of palm-oil mill effluent (POME) waste and the local conditions may favour choosing technologies for electricity generation from biogas.

We found that risks of possible leakage of methane emissions and unfiltered H_2S would occur on both pathways, as they are based on similar technological principles. Both risks were observed in the household biogas digesters (Devisscher *et al.*, 2017) and large-scale ones, such as the POME installations (Promnuan and O-Thong, 2017).

Risks associated only with household-level biogas

Risks associated to household biogas are different from the large scale for electricity, which will be explored in this section. For example, the barriers of collective biogas management at the community level would be unlikely to factor in the large-scale biogas-for-electricity pathway, since the plants would be run as a larger-scale operation with suitable management mechanisms and practices. Similarly, household-level gender discrepancy relates to unpaid work on the smallholding, and likewise, accessibility of the feedstock relates to shortages in the quantity of feedstock for household biogas.

Other risks in the household biogas pathway result mainly from the way the biogas distribution schemes have been designed and implemented. These risks include poor maintenance and infrastructure as well as bureaucratic issues among the potential beneficiaries. The lack of support for maintenance of household systems prevents the rural users from fully benefiting from their biogas installation. According to Berhe *et al.* (2017), the successful utilisation and maintenance of biogas infrastructure is commonly the joint responsibility of the owner and technical personnel. Clearly, when the capacity of the farmer (financial, skills, knowledge, etc.) is lower, the likelihood of this risk is higher. Increasing the capacity of these actors is necessary for mitigating the risks associated with poor maintenance and infrastructure.

Additionally, there are a number of risks on this pathway that may trigger the users to immediately revert back to conventional energy. For example, when biogas collective management is not well organised, it will create a higher likelihood of the users to seeking easily accessed firewood or even LPG. This is similar to other risks, such as poor maintenance, bureaucratic issues, inadequate incentives, and gender issues: where the farmers face limited options of biogas usage, there is a higher chance for them to seek a more familiar energy option.

Risks associated only with biogas-for-electricity generation

Risks exclusively attached to the biogas-for-electricity pathway concern the reliability of the electricity generation, as well as the increased electricity tariff, both of which were suggested by PLN. PLN actually purchased 1 MW of POME-based electricity in 2015 to distribute it through on-grid channels (Pasadena Engineering Indonesia, 2014). PLN was concerned about whether or not the electricity delivered to the consumers would be as stable as the existing commercial electricity generators (hydroelectricity, thermal, diesel, gas, and geothermal power). Moreover, the quality and quantity of POME could also have implications for the methane proportion in biogas (Yacob *et al.*, 2006). While the common efficiency of natural gas-generated electricity ranges around 38–42% (Breeze, 2014), biogas-generated electrical efficiency ranges between 28% and 40% (Firdaus *et al.*, 2017; Shi, 2011). Second, it was thought that the high operational costs would also contribute to the risks of on-grid electricity.

This might create difficulties in raising finances and it may also necessitate a higher tariff that could be damaging for consumer demand.

Overall, there were higher proportions of discussions on the risks related to household biogas than large-scale electrification, as household biogas was the starting point of the Indonesia narrative. For Bappenas, PLN, and MEMR, being the most familiar with the pathway of large-scale biogas for electricity, there was a knowledge discrepancy during the stakeholder dialogues, leading to a tendency to discuss this pathway less. This situation led stakeholders to focus more on the household biogas pathway. If the risks of the electrification pathway are less well understood, they may also have been underestimated here. In addition, differences in 'worldviews' as a part of the uncertainties centred on issues related to biogas installations at household and communal levels. There did not appear to be a lack of consensus around biogas for electrification that resulted in ambiguity and contributed to uncertainty. Stakeholders all thought that producing electricity from animal waste will increase electricity access in remote Indonesian islands, but all disagreed that upscaling biogas for village-wide power generation or commercial scale generation is realistic.

Conclusions

This research identified two pathways of biogas development to reduce emissions and to supply energy needs that could be potentially relevant in Indonesia, namely household biogas and large-scale biogas-for-electricity development pathways. Four sources of uncertainty were salient: (i) the unclear role of public and private sectors; (ii) the changing focus on each political term; (iii) lack of a national biogas target; and (iv) differences of perceptions among the stakeholders. These uncertainties give context to the risks perceived by stakeholders.

Of the risks identified, a higher number were classified as technological risks. The stakeholders tended to focus on the implementation risks (i.e. barriers) rather than the consequential risks (negative outcomes). Moreover, most agree that biogas development has not yet become a government priority. As a result, biogas is not being developed evenly across the country. This situation created a tendency to highlight and criticise current efforts in terms of the barriers experienced.

A comparison of pathways was conducted to increase awareness about what kind of risks to expect, and what kind of uncertainties might interact with and exacerbate these risks. Interestingly, both pathways include some common or similar risks, among them the suitability of technological choices, the possible technological constraints, and the cost of initial investment. Other risks exclusively pertained to one pathway or another. However, further studies on the electricity pathway are needed to identify a wider range of risk perceptions. The severity (potential impact) of each risk is not studied in this narrative.

The risks of the household biogas pathway may lead to difficulties for rural residents to continue using biogas, thus may result in unattainable renewable energy targets and prevent potential co-benefits such as protecting the environment and improving human health. On the other hand, the biogas-to-electricity pathway risks could create negative impacts on the economy, either commercially (on the market) or at the macro/national level. There-fore, to achieve the national target through these pathways, it is important to suppress the uncertainties at the higher level, or those mainly related to the governance scope, which could include, for example, making progress in for-mulating targets complemented with clear pathways and providing clarity in how biogas development should be perceived, valued, and approached. It is also recommended that policymakers implement supporting actions to miti-gate risks, such as strengthening the institutions that manage the national biogas development.

References

Asadullah, M. (2014). Barriers of commercial power generation using biomass gasifica-tion gas: A review. *Renewable and Sustainable Energy Reviews*, 29, 201–215. https://doi.org/10.1016/j.rser.2013.08.074.

Atteridge, A., Aung, M.T., Nugroho, A. (2018). *Contemporary coal dynamics in Indonesia* (SEI working paper). Stockholm Environment Institute, Stockholm.

Aven, T. (2012). The risk concept: historical and recent development trends. *Reliability Engineering & System Safety*, 99, 33–44. https://doi.org/10.1016/j.ress.2011.11.006.

Bedi, A.S., Sparrow, R., Tasciotti, L. (2017). The impact of a household biogas pro-gramme on energy use and expenditure in East Java. *Energy Economics* 68, 66–76. https://doi.org/10.1016/j.eneco.2017.09.006.

Berhe, T.G., Tesfahuney, R.G., Desta, G.A., Mekonnen, L.S. (2017). Biogas Plant Dis-tribution for Rural Household Sustainable Energy Supply in Africa. *Energy and Policy Research*, 4(1), 10–20. https://doi.org/10.1080/23815639.2017.1280432.

BIRU (2018). *Indonesia Domestic Biogas Programme: January–December 2017 (Annual Report)*. BIRU. Available at: www.biru.or.id/files/annual-report-2017.pdf.

Blenkinsopp, T., Coles, S., Kirwan, K. (2013). Renewable energy for rural communities in Maharashtra, India. *Energy Policy* 60, 192–199. https://doi.org/10.1016/j.enpol.2013.04.077.

BP (2018). *BP Statistical Review of World Energy* (No. 67). British Petroleum. Available at: www.bp.com/content/dam/bp/en/corporate/pdf/energy-economics/statistical-review/bp-stats-review-2018-full-report.pdf.

BPS (2017). *Statistik Nasional Indonesia*. Available at: www.bps.go.id/website/pdf_publikasi/Statistik-Indonesia-2016-_rev.pdf.

Breeze, P. (2014). Natural Gas-fired Gas Turbines and Combined Cycle Power Plants, in: *Power Generation Technologies* (2nd edn, 67–92). Newnes, Oxford, UK and Waltham, USA. Available at: www.elsevier.com/books/power-generation-technologies/breeze/978-0-08-098330-1#.

Brent, A.C., Kruger, W.J.L. (2009). Systems analyses and the sustainable transfer of renewable energy technologies: A focus on remote areas of Africa. *Renewable Energy* 34(7), 1774–1781. https://doi.org/10.1016/j.renene.2008.10.012.

Bruun, S., Jensen, L.S., Khanh Vu, V.T., Sommer, S. (2014). Small-scale household biogas digesters: An option for global warming mitigation or a potential climate bomb?

Renewable and Sustainable Energy Reviews, 33, 736–741. https://doi.org/10.1016/j.rser.2014.02.033.

Chaiprapat, S., Mardthing, R., Kantachote, D., Karnchanawong, S. (2011). Removal of hydrogen sulfide by complete aerobic oxidation in acidic biofiltration. *Process Biochemistry*, 46(1), 344–352. https://doi.org/10.1016/j.procbio.2010.09.007.

Cornot-Gandolphe, S. (2017). *Indonesia's electricity demand and the coal sector: export or meet domestic demand?* Oxford Institute for Energy Studies, Oxford, UK.

Climate Action Tracker (2018). Indonesia: Climate Action Tracker. Available at: https://climateactiontracker.org/countries/indonesia/ [accessed 30 July 2018].

Devisscher, T., Johnson, O., Suljada, T., Boessner, S., Taylor, R., Takama, T., … Yuwono, Y. (2017). *D6.2: Report on Social Discourse Analyses and Social Network Analyses* (Transitions pathways and risk analysis for climate change mitigation and adaptation strategies No. 642260). TRANSrisk.

Firdaus, N., Prasetyo, B.T., Sofyan, Y., Siregar, F. (2017). Part I of II: Palm Oil Mill Effluent (POME): Biogas Power Plant. *Distributed Generation & Alternative Energy Journal*, 32(2), 73–79. https://doi.org/10.1080/21563306.2017.11869110.

Friedrich, J., Ge, M., Damassa, T. (2015). Infographic: What Do Your Country's Emissions Look Like? Available at: www.wri.org/blog/2015/06/infographic-what-do-your-countrys-emissions-look [accessed 17 July 2017].

Gall, E.T., Carter, E.M., Matt Earnest, C., Stephens, B. (2013). Indoor Air Pollution in Developing Countries: Research and Implementation Needs for Improvements in Global Public Health. *American Journal of Public Health* 103(4), e67–e72. https://doi.org/10.2105/AJPH.2012.300955.

Government of Indonesia (2017). Regulation of the President of the Republic of Indonesia on the General Plan of National Energy, Pub. L. No. 22 (2017).

IEA (2007). Climate Policy, Uncertainty, and Investment Risk. OECD/IEA.

IEA (2014). National Energy Policy (Government Regulation No. 79/2014). Available at: www.iea.org/policiesandmeasures/pams/indonesia/name-140164-en.php [accessed 28 July 2017].

IEA (2015). Energy Policies Beyond IEA Countries: Indonesia 2015, in: R. Mulyana, *New and Renewable Energy and Energy Conservation Sector Strategy*. Presentation given to the International Energy Agency, Jakarta, March 2014.

International Institute for Sustainable Development (2018). *Missing the 23% Target: Roadblocks to the development of renewable energy in Indonesia* (GSI Report). Available at: www.iisd.org/sites/default/files/publications/roadblocks-indonesia-renewable-energy.pdf.

Kementerian ESDM (The Ministry of Energy and Mineral Resources) (2017). Buku Informasi Bioenergi. The book of bioenergy information. Direktorat Jenderal Energi Baru Terbarukan dan Konservasi Energi. Kementerian Energi dan Sumber Daya Mineral Republik Indonesia. Available at: https://lintasebtke.com/wp-content/uploads/2017/11/Buku-Informasi-Bioenergi.pdf.

Kusumaningrum, W.B., Munawar, S.S. (2014). Prospect of Bio-pellet as an Alternative Energy to Substitute Solid Fuel Based. *Energy Procedia*, 47, 303–309. https://doi.org/10.1016/j.egypro.2014.01.229.

Kuvarakul, T., Devi, T., Pratidina, A., Schweinfurth, A., Winarno, D., Sikumbang, I. (2014). *Renewable Energy Guidelines on Biomass and Biogas Power Project Development in Indonesia*. Jakarta: GIZ and ASEAN. Available at: www.fao.org/fileadmin/templates/rap/files/meetings/2014/140723-d2s2.indo.pdf.

Masini, A., Menichetti, E. (2012). The impact of behavioural factors in the renewable energy investment decision making process: Conceptual framework and empirical findings. *Energy Policy* 40, 28–38. https://doi.org/10.1016/j.enpol.2010.06.062.

OECD (2015). Survei Ekonomi OECD Indonesia. OECD. Available at: www.oecd.org/economy/Overview-Indonesia-2015-Bahasa.pdf.

Pasadena Engineering Indonesia (2014). Biogas: Palm Oil Mill Effluent/Waste water mills. Available at: http://pasadena-engineering.com/en/teknologi [accessed 8 June 2018].

Promnuan, K., & O-Thong, S. (2017). Efficiency Evaluation of Biofilter for Hydrogen Sulfide Removal from Palm Oil Mill Biogas. *Energy Procedia* 138, 564–568. https://doi.org/10.1016/j.egypro.2017.10.160.

Rahayu, A. S., Karsiwulan, D., Yuwono, H., Trisnawati, I., Mulyasari, S., Rahardjo, S., … Paramita, V. (2015). *Handbook POME-to-Biogas: Project Development in Indonesia* (2nd edn). Winrock International, Jakarta. Available at: http://winrock-indo.org/documents/CIRCLE%20Handbook%20%28English%29.pdf.

Republic of Indonesia (2015). *Indonesia First Biennial Update Report (BUR) under the United Nations Framework Convention on Climate Change.* Directorate General of Climate Change, Jakarta. Available at: https://unfccc.int/resource/docs/natc/idnbur1.pdf.

Secretariat General of National Energy Council (2016). Indonesia Energy Outlook 2016. Available at: www.ea-energianalyse.dk/reports/1635_ieo_2016.pdf.

Shi, C.Y. (2011). *Mass flow and energy efficiency of municipal wastewater treatment plants.* IWA Publishing, London.

Singh, R., Setiawan, A.D. (2013). Biomass energy policies and strategies: Harvesting potential in India and Indonesia. *Renewable and Sustainable Energy Reviews*, 22, 332–345. https://doi.org/10.1016/j.rser.2013.01.043.

Siregar, U.J., Narendra, B.H., Suryana, J., Siregar, C.A., Weston, C. (2017). Evaluation on community tree plantations as sustainable source for rural bioenergy in Indonesia. *IOP Conference Series: Earth and Environmental Science*, 65, 012019. https://doi.org/10.1088/1755-1315/65/1/012019.

Urmee, T., Md, A. (2016). Social, cultural and political dimensions of off-grid renewable energy programs in developing countries. *Renewable Energy* 93, 159–167. https://doi.org/10.1016/j.renene.2016.02.040.

Widodo, T.W., Rahmarestia, E. (2008). Current status of bioenergy development in Indonesia. *Reg. Forum Bioenergy Sect. Dev. Challenges, Oppor. W. Forw.*, 49–66. Available at: www.researchgate.net/profile/Elita_Widjaya/publication/237290589_Current_Status_of_Bioenergy_Development_in_Indonesia/links/5ab9b8880f7e9b68ef533a8e/Current-Status-of-Bioenergy-Development-in-Indonesia.pdf.

World Bank (2018, 27 March). Indonesia Continues to Build on Solid Economic Growth. Available at: www.worldbank.org/en/news/press-release/2018/03/27/indonesia-continues-to-build-on-solid-economic-growth [accessed 17 July 2018].

Yacob, S., Ali Hassan, M., Shirai, Y., Wakisaka, M., Subash, S. (2006). Baseline study of methane emission from anaerobic ponds of palm oil mill effluent treatment. *Science of The Total Environment* 366(1), 187–196. https://doi.org/10.1016/j.scitotenv.2005.07.003.

13 Kenya

Risks and uncertainties around low-carbon energy pathways

Oliver W. Johnson, Hannah Wanjiru, Mbeo Ogeya, and Francis X. Johnson

Introduction

Like many countries in sub-Saharan Africa, Kenya has high development ambitions, aiming to become a middle-income country by 2030 (Government of Kenya, 2007). These ambitions are based upon a low-carbon, climate-resilient development pathway, as set out in Kenya's Nationally Determined Contributions (NDC) to global climate change mitigation under the UNFCCC (Ministry of Environment and Natural Resources, 2015). These ambitions depend on rapid expansion of the energy sector to increase the access, security, and affordability of energy service provision (Ministry of Energy and Petroleum, 2016). There are multiple, complementary pathways that Kenya might pursue to achieve these goals. In this chapter, we focus on two of those pathways: expansion of geothermal power development and development of greater sustainable charcoal production.

In the power sector, the country's abundant renewable energy resources offer significant opportunities for pursuing low-carbon development pathways. Geothermal in particular is poised to play an important role in such pathways. Having already moved from niche technology to the mainstream, geothermal power development now sits at an important threshold: after years of public-led investment in development of the geothermal sector, Kenya is seeking more private-sector-led expansion of its vast remaining geothermal resources (Musembi, 2014). As a world leader on geothermal energy, the Kenyan experience offers valuable lessons for research and has already been analysed in global comparisons of sustainability indicators (Shortall, Davidsdottir, and Axelsson, 2015).

At the same time, Kenya is seeking to modernise its cooking sector, which remains dominated by traditional biomass fuels with significant negative impacts on land, ecology, and emissions. Increasing urbanisation will have significant implications for forest resources if charcoal continues to remain the most affordable and accessible urban cooking fuel. However, more research is needed to better understand the complexities and uncertainties around greenhouse gas (GHG) emissions in land-use and biomass sectors (Dalla Longa and van der Zwaan, 2017). Through a mix of regulation, promotion of improved kilns and cookstoves, and support of alternative fuels, Kenya seeks a dual approach of increasing the sustainability of charcoal production, trade and consumption, and providing opportunities for fuel switching (Ministry of Environment, Water and Natural Resources, 2013; Wanjiru, Nyambane, and Omedo, 2016).

In this chapter, we explore the risks and uncertainties associated with expansion of geothermal power development and sustainable charcoal production and trade, framing them within the discourse around Kenya's energy future. We first situate the development of geothermal and sustainable charcoal sectors within the context of Kenya's low-carbon transition pathway and the country's changing political landscape. We then present analyses of risks and uncertainties around further development within both sectors, followed by a discussion of their implications. We conclude with policy recommendations for the two sectors.

Research process and methods

To explore the risks and uncertainties around developing geothermal power generation and sustainable charcoal production and trade in Kenya, we analysed the technological innovation system (see Bergek *et al.*, 2008; Hekkert *et al.*, 2007), including the market system and actors. We used stakeholder attribute matrices, stakeholder engagement in the form of interviews and focus group discussions (FGDs), and field observations to identify the risks and uncertainties faced by different actors across the market system. We undertook 17 semi-structured interviews and three FGDs with geothermal-sector stakeholders, including three national government officials, one county government official, three employees of state-owned utilities, two representatives of independent power producers, one representative of the regulatory authority, two members of research institutions, five representatives of development partners, and seven members of the local community in the Olkaria region. We also undertook seven semi-structured interviews and targeted discussions during two workshops with charcoal-sector stakeholders, including six national government officials, two county government officials, representatives of three charcoal producer associations, representatives of one charcoal transporter group, four members of research institutions, and two representatives of development partners. Data collection took place over the course of June 2016 to May 2018.

Low-carbon energy pathways for Kenya

Kenya is one of East Africa's major economies, with a population of almost 50 million and a GDP of roughly US$75 billion.[1] The country's high dependence on natural resources makes its GDP very sensitive to impacts of climate change on the natural environment (Ministry of Environment and Natural Resources, 2016). In 2007, the government established Vision 2030, its blueprint for becoming a newly industrialising, middle-income country providing a high-quality life for all its citizens by 2030 (Government of Kenya, 2007). The notion of a green economy resonates through Vision 2030, merging the imperatives of contributing to global climate change mitigation, meeting the increasing energy demands of a growing population and economy, sustainably managing the country's valuable natural resources, and enhancing climate resilience.

Committing to climate change mitigation and adaptation

In 2015, Kenya submitted its NDC to the UNFCCC, noting that land use, land-use change, forestry, and agriculture contributed 75% of the total GHG emissions in 2010 (Ministry of Environment and Natural Resources, 2015). By the year 2030, Kenya's GHG emissions under a business-as-usual scenario – excluding future exploitation in the extractive industry – are estimated at 143 $MtCO_2e$,[2] based on the per capita emissions of about 1.26 $MtCO_2e$ (Ministry of Environment and Natural Resources, 2015). Therefore, the country has announced mitigation and adaptation actions to abate its GHG emissions by 30%. To achieve a low-carbon, climate-resilient development pathway, some of the mitigation activities relate to promoting clean energy technologies to reduce overreliance on wood fuels; achieving a tree cover of at least 10% of the land area; and expansion of renewable sources of energy. Geothermal power development and sustainable charcoal production and trade both form important elements of Kenya's mitigation priorities and, as such, both were the focus of two recent proposed Nationally Appropriate Mitigation Actions for Kenya submitted to the UNFCCC (Falzon et al., 2014; Wanjiru et al., 2016).

Meanwhile, as with many developing countries in Africa that are vulnerable to climate change, adaptation measures will continue to receive support, including 'climate proofing' infrastructure as well as supporting innovation and development of appropriate technologies that promote climate-resilient development. These measures have come out of Kenya's consecutive strategies and plans for addressing climate change, starting with the National Climate Change Response Strategy (NCCRS) developed in 2010, followed by the National Climate Change Action Plan (NCCAP) launched in 2013 and revised in 2018, the Climate Change Act passed in 2016, and the National Adaptation Plan (NAP) which was finalised in 2016 (Government of Kenya, 2010a, 2013, 2016a, 2016b, 2018a).

Powering the nation and fuelling its cities

As shown in Figure 13.1, renewable energy has always played a major role in Kenya's electricity supply. Hydro has typically dominated the electricity mix, but over the last 15 years geothermal has taken an increasing share and wind has started to become prominent. By 2018, hydro, geothermal, and wind made up 35.3%, 27.9%, and 1.1% respectively of the country's 2,333 MW of installed grid-connected electricity (Kenya Power, 2018).

As electricity access expands, the middle class grows, and industrial activity rises, major demand increases are forecast (Government of Kenya, 2011, 2018b).

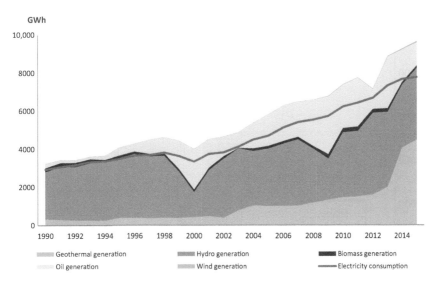

Figure 13.1 Electricity generation and consumption in Kenya, 1990–2015.

Source: www.iea.org.

Since 2011, a number of plans have been established to meet this demand, the latest being the updated Least Cost Power Development Plan (LCPDP) 2017–2037, published in 2018. In it, the long-term target for generation expansion is around 7213 MW by 2030 and 9932 MW by 2037, as shown in Figure 13.2 (Government of Kenya, 2018b). Also in 2018, the government announced its third Medium-Term Plan (2018–2022) – its economic development vehicle anchored within the Vision 2030 – giving great priority to expanding the renewable energy sector.[3] Figure 13.2 shows how renewable energy is expected to feature in electricity generation expansion over the next 20 years.

Both the LCPDP 2017–2037 and the third Medium Term Plan envisage a four-fold expansion of geothermal power generation from 650 MW to around 2500 MW in 20 years. Abundant, low carbon, and climate resilient, geothermal power is an attractive resource with potential for additional heat applications in industry. And over the past four decades, considerable technical expertise in geothermal has been established within the country's state-owned utilities and ancillary services. However, attracting the private investments needed to develop the country's geothermal resources at a swifter pace remains a challenge.

Meanwhile, in the cooking sector, a rising and increasingly urbanising population is demanding more charcoal, exerting increasing pressure on forests, farmlands, and community rangelands from where it is sourced. Charcoal – produced in kilns by carbonising wood by pyrolysis – meets the cooking energy needs of over 80% of Kenya's urban population (Wanleys Consultancy Services, 2013).

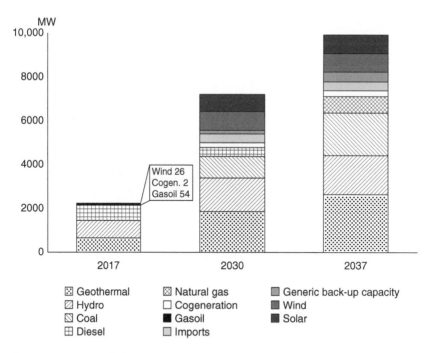

Figure 13.2 Projected electricity generation in Kenya, 2017–2037.

Source: www.iea.org.

But the growing gap between supply and demand of the commodity will only expand unless action is taken. Imports from neighbouring countries meet some of this gap, but not all. Ecological impacts and land degradation from the remaining unsustainable charcoal production in Kenya add to its GHG emissions and threaten future livelihoods due to declining yields, decreased biodiversity, and other impacts (Kiruki *et al.*, 2017; Ndegwa *et al.*, 2016a).

Putting the Kenyan charcoal sector on a more sustainable pathway calls for innovative approaches across the charcoal market chain that can improve efficiency in harvesting, production, transport, distribution, and consumption. Yet doing so is not easy: the sector remains informal with little recognition in national economic reporting despite employing hundreds of thousands of people and generating hundreds of millions of US dollars (Njenga *et al.*, 2013). Although it is an important source of livelihood for some, the economic returns tend to be concentrated among larger producers and wholesale traders, while small-scale producers may effectively be trapped in poverty (Ndegwa *et al.*, 2016b; Zulu and Richardson, 2013).

Navigating the changing political landscape

As Figure 13.3 shows, there is a range of legislation, policies, and strategies influencing the development of the geothermal and charcoal sectors in Kenya, some broad in scope and others with a sectoral focus. But in recent years, decisions around how to manage geothermal power generation and charcoal production and trade have also been heavily influenced by the changing political landscape associated with the devolved government system that was established in the wake of a new Kenyan constitution in 2010 (Government of Kenya, 2010b). Debate continues over how governance of the energy sector at county and central governments will be managed in practice. The devolved system in Kenya is still new, hence a lot of learning and adaptation still needs to take place before effective means of ensuring citizen participation are established.

With energy planning and development mandates, county governments have a substantial role to play with regard to shaping energy development priorities and politics according to their local resources (Johnson *et al.*, 2016). For instance, most geothermal steam fields lie within the Rift Valley – an area spreading across Turkana, Baringo, Nakuru, and Kajiado counties. Local governments in these counties want a role in decision making over geothermal development in their constituencies to embrace its benefits, rather than risk disruptions in their county and local community (Matara and Sayagie, 2018). Meanwhile, some charcoal production hotspot areas, such as Tharaka-Nithi, Kitui, Narok, Kajiado, and Kwale counties, have already developed regulations to manage how their woody biomass resources are used and preserved (Wanjiru *et al.*, 2016). However, county-level budgets generally support roads or other infrastructure rather than energy provision. Large-scale energy infrastructure remains outside the purview of county governments, while small-scale solutions and household use of biomass energy receive little political attention (Johnson *et al.*, 2016).

Much hinges on the 2017 Energy Bill – first put forward in 2015 and currently under consideration by the Parliament – which will give legal clarity as to what local-level governance would mean within the counties when it comes to energy issues (Government of Kenya, 2017). For example, each county government is expected to develop a county energy master plan that will be used by the Cabinet Secretary of the Ministry of Energy and Petroleum to formulate an integrated national energy master plan for purposes of national energy planning. In addition, county governments will have the power to enforce certain provisions for efficient use of energy and its conservation, to undertake inspections, and to issue directions, all in relation to national energy laws and provisions. Furthermore, for geothermal projects, county governments will receive 20% of the royalties from geothermal power produced in their jurisdictions, and the local community shall receive 5% of the royalties through a community trust fund.

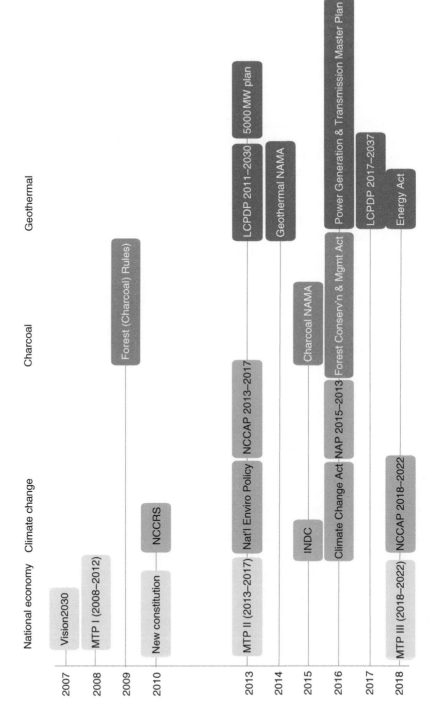

Figure 13.3 Range of legislation, policy, and strategy governing Kenya's energy pathways.

Risks and uncertainty in Kenya's energy pathways

Geothermal power development and sustainable charcoal production and trade both form important energy pathways in Kenya's quest to become a middle-income country based on a climate-resilient green economy. But implementation and consequences of these pathways are by no means certain: understanding the risks and uncertainties around these pathways is essential to overcoming barriers and minimising negative impacts.

Geothermal power development

The technological innovation system life cycle for geothermal encompasses roughly six phases, taking place over up to a decade (see ESMAP, 2012; Ng'ang'a, 2005). Geothermal development starts with geo-exploration through surface studies followed by exploratory drilling, a practice that involves drilling three to six narrow wells to about 2000–3000 metres. Once the resource is proven viable, around a dozen production wells are drilled to extract steam, and a system of pipes is constructed to gather the steam at one location and to then reinject it back into the steam field reservoir. The gathered steam is most commonly used indirectly in a steam turbine for power generation. It can also be used directly for a range of heat applications, such as spas, district heating, and industrial and smaller-scale processes requiring heat. In the case of steam turbine power generation, power is typically transmitted and distributed through the national grid to residential, commercial, and industrial end users. In both cases, steam field management is crucial to ensure the resource is not depleted and that hazardous chemicals in the steam are properly managed. Decommissioning of geothermal steam fields and power plants has yet to be experienced in Kenya.

Historical perspective on geothermal

Kenya's geothermal resource is located within the country's Rift Valley, with recent estimates suggesting a resource potential of between 7000 MW and 10,000 MW spread over 14 sites (Ngugi, 2012). Exploration in the Olkaria steam field in the late 1960s to mid-1970s by the state-owned Kenya Power Company Limited and supported by the UNDP led to the drilling of production wells in the Olkaria I block and commissioning of a 15 MW geothermal power plant in 1981. Drilling continued, with up to 20 wells added by 1985, and two additional 15 MW power plants were commissioned in 1982 and 1985 (Omenda and Simiyu, 2015; Riaroh and Okoth, 1994; Simiyu, 2008).

Reform of the power sector in 1997 led to the unbundling of Kenya Power Company Limited into two entities: Kenya Power and Lighting Company (KPLC) – later rebranded as Kenya Power – responsible for transmission and distribution, and Kenya Electricity Generating Company (KenGen) responsible for generation (Kapika and Eberhard, 2013; Karekezi and Mutiso, 2000). In its new form, KenGen remained in control of the Olkaria I block and began drilling in Olkaria

II. In the meantime, the first private-sector concession was awarded to OrPower4 in 1998 to explore and develop Olkaria III. Additional Olkaria II steam turbine units were commissioned in 2003 and in 2007; the 140 MW Olkaria IV was commissioned in 2010; and current work is ongoing to develop another 140 MW in Olkaria V (Kenya Power, 2018; Ngugi, 2012; Omenda and Simiyu, 2015). Since serious geothermal exploration first began 40 years ago, geothermal has evolved from niche technology and resource to being a major contributor to the national electricity mix, with an installed capacity of 652 MW providing almost half of Kenya's power (Kenya Power, 2018).

Geothermal capacity is projected to reach over 5500 MW by 2030, but only if greater private-sector involvement can be achieved (Ngugi, 2012; Omenda and Simiyu, 2015). To help accelerate geothermal development, the government established the Geothermal Development Company (GDC) in 2009, with a mandate to carry out rapid exploration and development of geothermal over the next 20 years, encouraging further private-sector-led expansion in geothermal power generation, and removing the high risks associated with expensive exploratory drilling (Ngugi, 2012). A decade later, acceleration has been limited. GDC is currently developing a geothermal field in Menengai providing steam sales to three independent power producers (IPPs), but the project has experienced delays related to finalising the steam sales agreement and getting government letters of support, both of which are essential to convince investors that financial and political risks are manageable. Other fields are promising but much hinges on progress in Menengai (Ministry of Energy and Petroleum, 2013; Ngugi, 2012).

Implementation risks

Geothermal development faces a range of barriers or potential risks to implementation (Figure 13.4). In terms of economic feasibility, geothermal development on 'greenfield' sites – where no previous development has taken place – requires considerable upfront investment. One exploration well costs over US$1 million to drill, and three wells are required simply to prove the resource. This high investment is prohibitively risky for both private companies looking to ensure a return on investment and state-owned utilities with limited budgets. In Olkaria, representatives from KenGen and OrPower admit they have been very lucky to find steam so easily and that the quality of steam has remained consistent for so long. This might not be the case elsewhere in the Rift Valley, and delays faced by private companies in Akiira and Longonot show the difficulty in finding investors patient enough to finance additional exploration. Stakeholders highlight that GDC was created precisely to bear this risk on behalf of the private sector, undertaking exploration and steam field development in greenfield sites and selling the steam to IPPs, which invest in power generation only.

But even once the resource is proven, the financial risk does not disappear. Typical costs for a 20 MW geothermal power plant – including these production

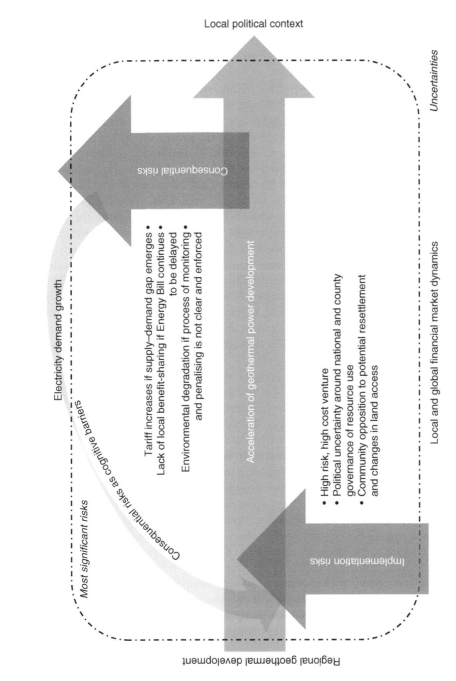

Figure 13.4 Main uncertainties, implementation, and consequential risks associated with the geothermal power development in Kenya.

wells, the steam gathering system, and steam turbine technology – can reach almost US$80 million (GEOCOM, 2015). Recent slow growth in demand due to limited economic and industrial development[4] has led to growing concern among government and Kenya Power officials that they will not have enough customers to be able to pay for the electricity that the company has agreed to purchase.

This economic risk associated with inability of the 'offtaker' – Kenya Power – to pay for electricity it has contractually agreed to purchase is a key political and institutional risk facing the geothermal sector. This may, in turn, greatly increase the cost of borrowing capital for investment. In the private sector, investors with a high tolerance for risk may be more amenable to investing in geothermal, but they typically require strong guarantees before they are willing to lend to green-field project developers or IPPs. In Menengai, GDC developed the geothermal field and will sell steam to three IPPs. To mitigate investment risks, the IPPs have a project and steam sales agreement with GDC to guarantee the steam they will receive, and a power purchase agreement with Kenya Power to buy the power they produce. But delays in closing financing for the IPPs have continued as letters of support from the government have been slow in forthcoming, leaving some polit-ical risks unresolved. Investors, private developers, and government stakeholders all appeared to have very different perspectives on who should bear which risk. As such, geothermal remains dominated by grants and concessional loans (high interest and long tenors) from development finance institutions, such as the Euro-pean Investment Bank, the KfW Development Bank, the World Bank and the Japan International Cooperation Agency (JICA).

Another political risk from the perspective of many different stakeholders is the distribution of responsibility for energy planning and project approval between the national and the county governments. County government repre-sentatives felt that too much national control was a risk to their ability to manage their own affairs and ensure that the voice of county citizens was repres-ented. National government representatives viewed added bureaucracy and potential for political manoeuvring as a risk to project development and approval, which many deemed was already overly convoluted. Meanwhile, private developers sat on the fence, appreciating the role county government could play in managing local issues but remaining wary that increased levels of bureaucracy might lead to increased avenues for corruption, which already per-vades so much of the Kenyan economy. The 2017 Energy Bill, still awaiting final approval, may do much to clarify the allocation responsibilities among national and county governments; however, limited capacity at the county level, and ambiguities in the details, will take years to resolve.

The final risk associated with achieving further geothermal development is community opposition to construction both at the geothermal site and for the associated transmission lines to connect these sites with distant demand centres, such as major cities and industrial areas. In the face of relocation – which occurred in 2010 to facilitate development of the Olkaria IV power plant – or other restric-tions on land use, communities are understandably often resistant to geothermal development. Geothermal is not special in this regard: many other large energy infrastructure projects, including various wind power and transmission line projects, have faced community opposition, spanning from written complaints to roadblocks and open protests. Government, state-owned utilities, and private project developers take community concerns seriously. For example, GDC has enlisted support from the New Zealand government in how to manage engagement with Indigenous groups on infrastructure development and land issues: the two countries have had some common and comparable experiences (Shortall, Davidsdottir, and Axelsson, 2015). Meanwhile, development partners, such as the World Bank, place considerable pressure on their projects to minimise social risks and show compliance with international standards such as the Equator Principles.[5] Efforts to ensure the local community shares some of the benefits of infrastructure development typically include compensation, relocation to upgraded housing and additional community facilities, and job opportunities for unskilled labour in project construction and post-construction security. But the complex financing arrangements of geothermal projects often impede or limit on-the-ground implementation of international standards of development finance (Ole Koissaba, 2018). The benefits are often incomparable to the losses or unevenly distributed: for instance, it might be impossible to weigh improved access to a medical clinic with loss of fertile land for grazing livestock.

Consequential risks

Geothermal brings many benefits in terms of a low-carbon, climate-resilient source of electricity, with the potential for high output into the national grid. But there is also a risk of a range of adverse consequences (refer back to Figure 13.4). With careful management of steam reservoirs – as is currently the case in Olkaria – geothermal is a renewable source with little financial burden once initial capital costs are paid back. This results in competitive electricity tariffs that can help to reduce the consumer cost of electricity, as already experienced in Kenya. However, if supply does not match demand, then higher unit costs may prevent Kenya Power from lowering tariffs or may require them to raise tariffs to cover the costs of meeting obligations in power purchase agreements. This is a risk for the consumers who may have to pay more, for Kenya Power who may lose money and reputation, for the government and taxpayers who may have to subsidise Kenya Power, and for the energy regulator who may find it impossible to balance cost recovery with affordable tariffs.

Risks of negative social impacts of geothermal development – beyond those associated with higher electricity costs – largely revolve around negative impacts on livelihoods and the culture of local communities in the vicinity of steam field and power plant infrastructure. These negative impacts may perpetuate existing social inequalities and the marginalisation of traditional societies. While the urban middle class, industries, and manufacturing businesses benefit from cheaper and more reliable electricity, access of local communities to training and skilled employment at geothermal sites might remain limited. The new

Energy Bill currently awaiting final approval will establish a community fund for development activities, which is likely to help significantly to achieve greater benefit sharing (Government of Kenya, 2017). But until this comes into being and is proven to help, it remains unclear whether devolved government – with its added layer of political dynamics – will mitigate or exacerbate social impacts.

The environmental risks of upscaled geothermal development include contamination from poor handling of toxic chemicals in the steam; withdrawal of water from lakes, rivers, and wells beyond their capacity; and degradation and disruption to natural habitats and migratory routes of wildlife inside and outside protected areas (see Kubo, 2003; Mariita, 2002; Mwangi, 2005; Ogola, Davidsdottir, and Fridleifsson, 2012). These risks are largely the concern of conservation groups and others dependent on clean and available land and water resources. They can be – and often are – allayed by enforcing extensive environmental impact assessments and strong risk mitigation measures, such as controlled reinjection of steam into reservoirs; regulated water withdrawal; wildlife-friendly steam piping designs; use of noise-reduction technology; and cautious management of toxic chemicals using the latest technology and processes. However, non-compliance can result in severe impacts. The situation calls for extra measures to enforce the set regulations, and perhaps giving more positive visibility to those who pursue best practices.

Sustainable charcoal production and trade

The charcoal technological innovation system life cycle encompasses six phases. Charcoal production begins with harvesting woody biomass from communal land, government forest, and private land (Njenga *et al.*, 2013). The woody biomass is then carbonised by pyrolysis in a kiln to produce a certain charcoal, with typical kiln 'efficiencies' – the ratio of charcoal mass output to dry wood mass input – ranging from 10% to 30% (Bailis, 2009; Ministry of Environment, Water and Natural Resources, 2013). Charcoal is then transported in 50–90 kg sacks from production sites to urban and peri-urban demand sites. From there it is then distributed to consumers in a range of sizes, from whole sacks to 20-litre buckets to two-litre tins. Efficiency in final use of charcoal for cooking depends on the stove technology that consumers own and prefer to use. Some entrepreneurs have started to make charcoal briquettes from charcoal dust created during production, transportation, and distribution – amounting to roughly 25% of total original charcoal volume.

Historical perspective on charcoal

The increase of charcoal use in Kenya is largely a function of two factors: urban population growth and limited switching to alternative fuels. Between 1960 and 2017, Kenya's population rose over sixfold, and the proportion of the population living in urban areas more than tripled.[6] In the 1980s, charcoal was used by 50% of the urban population (O'Keefe, Raskin, and Bernow, 1984) but by 2002 this figure was reported to have reached 82% (Kamfor Company, 2002). Although liquefied petroleum gas (LPG), electricity, and pellets burned in gasifier stoves are available as alternative fuels for cooking in urban centres, they have remained the preserve of high-income households only. As such, charcoal continues to be a major source of cooking fuel for urban households across all income levels (Dalberg Advisors, 2018; Kojima, Bacon, and Zhou, 2011).

Rapidly rising demand for charcoal has led to a widening supply–demand gap resulting in unsustainable charcoal production. Current demand, estimated at 16.3 million m^3, is far above the current supply of 7.4 million m^3 (Wanleys Consultancy Services, 2013). And by 2032, demand is expected to increase by 17.8% while supply is expected to increase by only 16.8%, widening the supply-demand gap from 8.9 million cubic metres (m^3) to 10.6 million cubic metres. The charcoal market chain is a vital source of employment for over 500,000 people and generates over US$427 million, yet it is barely recognised in formal national economic reporting and forecasting (Njenga *et al.*, 2013).

Charcoal conservation efforts started in earnest in mid-1990s with the promotion of a more efficient charcoal stove – the Kenya Ceramic Jiko – by GTZ, the Kenya government, universities, and other development partners (Karekezi and Turyareeba, 1995; Tigabu, 2017). But efforts to make the charcoal sector sustainable – i.e. ensuring the charcoal that reaches homes and businesses is produced in a way that does not contribute to degradation of forests, lands, and ecosystems – has only become a focal point for action in the last decade. A legal framework regulating the production of charcoal, the Forest (Charcoal) Rules, was established in 2009 (Government of Kenya, 2009). Community forest associations or other common interest groups that register as formal charcoal producer associations with the Kenya Forest Service receive a registration certificate. Transporters buying from these registered charcoal producer associations receive a certificate of origin, which they present to their local Kenya Forest Service office to obtain a charcoal movement permit costing 500–1,000 KES (US$5–10) per trip.

In 2013, the National Environment Policy highlighted charcoal burning as a major threat for arid and semi-arid land and national forest degradation (Ministry of Environment and Natural Resources, 2013). In parallel, the National Climate Change Action Plan 2013–2017 highlighted charcoal production as a main contributor to GHG emission in Kenya and proposed the introduction of more efficient kilns as the most significant low-carbon opportunity to reduce emissions; these proposals were also included in the recently revised National Climate Change Action Plan 2018–2022 (Government of Kenya, 2013, 2018a). Meanwhile, a similar observation was reported in Kenya's Second National Communication to the UNFCCC on its NDC in 2015, with sustainable charcoal production considered to be a key part of Kenya's commitment to global climate change mitigation (Ministry of Environment and Natural Resources, 2015).

Implementation risks

Figure 13.5 highlights the risks to implementation of sustainable charcoal. The financing risks relate to covering the cost of forest management interventions and supporting the purchase of improved technology for charcoal production (kilns) and charcoal consumption (cookstoves). Charcoal production is widely considered to be a significant cause of deforestation and forest degradation, although the precise causal pathway is distorted by links to other drivers of forest degradation such as timber extraction, grazing of livestock, and clearing of forests to make way for crop production (Bailis *et al.*, 2017; Hosonuma *et al.*, 2012). According to many stakeholders, there is almost no financing available to fund the farm forestry and reforestation interventions necessary to establish a sustainable supply of biomass for charcoal production and to maintain forest cover. Meanwhile, charcoal producer associations note that efficient charcoal production technologies present a considerable financial expenditure for their members, who typically earn a low and unstable income and have little formal access to credit. Innovative financing mechanisms were widely considered vital to facilitating purchase of these improved production technologies, but stakeholders acknowledged that making formal lending solutions work within a largely informal sector presents a considerable obstacle. Those working on the charcoal demand side noted greater success in consumer-financing schemes for efficient charcoal consumption technologies, such as improved cookstoves, but warned of a distribution market marred by a wide variation in product quality. The irony of these financial risks is that, if the sector were streamlined, the government would retain about US$60 million with a 16% VAT rate, which potentially could be reinvested into the sector and thus used to manage financial risk (Ministry of Environment, Water and Natural Resources, 2013; Mutimba and Barasa, 2005).

Another risk is weak enforcement of the formal permitting system under the 2009 Forest (Charcoal) Rules. The system has done little to disincentivise production, transport, and use of charcoal from unsustainable sources. Kenya's informal system of bribes – so well-established that producers, transporters, and wholesalers have come to view it as an acceptable component of the charcoal trade – exacerbates the situation. Meanwhile, the formal permitting system is new, and the compliance requirements are often misunderstood by the traffic police and Kenya Forest Service officers tasked with verifying the validity of all movement permits. Officers are often individual beneficiaries of bribes and thus may have little incentive to enforce a formal permit system that instead benefits the local or national government. Since devolved county governments were created in 2013, counties with charcoal production hotspots – such as Kitui, Narok, and Kajiado counties – have started to establish and enforce their own regulations with which local charcoal producer associations and transporters have to comply. It is yet to be seen if they will prove more effective and enforceable than the national regulations.

The third risk is associated with competition from alternative cooking fuels. While a sustainable charcoal sector is an attractive proposition to some, others

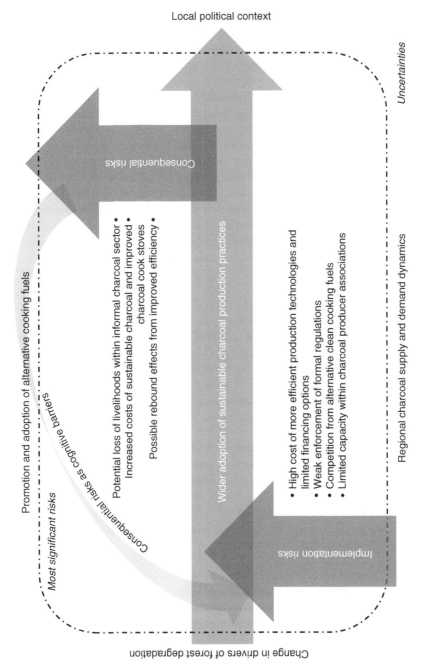

Local political context

Consequential risks

Implementation risks

Promotion and adoption of alternative cooking fuels

Most significant risks

Consequential risks as cognitive barriers

Potential loss of livelihoods within informal charcoal sector •
Increased costs of sustainable charcoal and improved charcoal cook stoves •
Possible rebound effects from improved efficiency •

Wider adoption of sustainable charcoal production practices

• High cost of more efficient production technologies and limited financing options
• Weak enforcement of formal regulations
• Competition from alternative clean cooking fuels
• Limited capacity within charcoal producer associations

Regional charcoal supply and demand dynamics

Uncertainties

Change in drivers of forest degradation

Figure 13.5 Main uncertainties, implementation, and consequential risks associated with sustainable charcoal production and trade in Kenya.

within the energy, public health, and environment sectors view charcoal as a dirty fuel that should be fully replaced by much cleaner alternatives, such as LPG, ethanol, and biomass pellets burned in gasifier stoves. Interventions to make these alternative fuels more available and affordable, particularly in urban areas, may pose a significant risk to sustainable charcoal production activities. If alternatives capture a sizeable share of the urban household energy market, it might reduce demand for charcoal, with potentially negative implications on charcoal-linked employment and livelihoods.

The final risk to greater pursuit of sustainable charcoal production is limited capacity within charcoal producer associations to ensure compliance with formal regulations. These regulations require registered charcoal producer associations to establish a written constitution, develop a conservation and reforestation plan, and document the kiln technologies and tree species they use to make charcoal. But tracking biomass resources and developing forest management plans require significant expertise that many members of these associations lack. Meanwhile, creating a constitution that all members agree upon is a challenge, especially as many fear formal regulation of informal and unregulated livelihoods.

Consequential risks

Despite its potential benefits, greater sustainable charcoal production poses some consequential risks (refer back to Figure 13.5). Many stakeholders perceive a streamlined increased market price of charcoal as a potential consequential risk to the long-term viability of sustainable charcoal, with actors along the market chain likely to pass along to consumers the additional costs of doing business formally. This includes transporters who may also continue to include the cost of bribes due to strong enforcement of informal practices and weak enforcement of formal rules.

There is also significant potential for loss of livelihoods if stricter enforcement of sustainable charcoal production takes place. The half a million people actively engaged in the charcoal market chain support an estimated two million dependants (Mutimba and Barasa, 2005). While the sustainable practices are meant to support more livelihoods in the long run with minimal environmental impacts, some stakeholders felt that only established businesses or community organised groups are likely to benefit.

Finally, development of more efficient charcoal kilns and cookstoves might lead to a rebound effect whereby more efficient use leads to fuel savings that in turn increase fuel use. For example, people might use charcoal to boil water or cook more often than they previously did (see Mwampamba et al., 2013). Reliance on biomass may increase, but scant research explores how greater efficiency in production and consumption technologies will affect supply and demand trends.

Synthesis

Energy pathways

Geothermal power development and sustainable charcoal production form complementary but very different pathways which can contribute to Kenya's vision of a low-carbon, climate-resilient future (Table 13.1). Geothermal power generation is a large-scale industry, boasting only a few main actors that are part of, or fit well into, the existing centralised electricity system. Transmission and distribution of geothermal power is similarly large in scale with limited actors. On the other hand, sustainable charcoal production is a cottage industry comprising a myriad of decentralised, small-scale actors. Transport, wholesale, and distribution is similarly small in scale and undertaken by thousands of different actors.

The technological capabilities required for upscaling geothermal power development lie partially within Kenya – where there has been significant accumulation of expertise and process innovation – and partially within foreign firms that possess the most advanced technology used in drilling and power generation and the most sophisticated knowledge of steam reservoir modelling and steam field management. The technological capabilities required to upscale sustainable charcoal production, on the other hand, can nearly all be found in Kenya. Local manufacturers of more efficient (and appropriate) charcoal kilns exist, although their products are not necessarily widespread. And the knowledge required to

Table 13.1 Selected aspects of energy pathways

Component	Geothermal power generation	Sustainable charcoal production
Structure	Centralised	Decentralised
End use	Electricity, industrial heating	Cooking
Investment needs	High capital cost by big investors	Local actors/small capital cost, although relatively high
Sector structure	Top-down	Bottom-up
Technological capabilities	Local innovation in processes but foreign technology	Local innovation in technology
Implementation risks	High cost and high chance of failure Political uncertainty around sector Public opposition	Technology costs Non-harmonised policy and weak enforcement Competing alternative fuels Limited capacity of charcoal producer associations
Consequential risks	High electricity costs from limited demand Limited benefit-sharing Environmental impact	High charcoal and cookstove costs Loss of livelihoods Potential rebound effects

develop and manage forests in a sustainable manner certainly exists, although it is not necessarily in the hands of those who need it most, namely the owners or users of land where charcoal is produced. Both pathways face considerable challenges in obtaining the finance needed for upscaling. Both have trouble accessing capital due to perceived risks.

Distribution of risks

It is not easy to identify risks associated with implementation and consequences of geothermal power development and sustainable charcoal production – and neither are the distribution and comparisons of those risks. The fairly rudimentary framework used in this study, in which we separated risks associated with each energy pathway into implementation and consequential risks, has limitations given that different actors clearly frame risks and benefits in different ways. Indeed, it is possible to separate the risks according to how they are distributed between private economic interests, duty bearers, and rights holders.

Within geothermal, many of the financial risks fall upon private investors and project developers seeking to invest debt and equity into exploration, drilling, and power generation ventures. Investors and developers are typically willing to shoulder risks that are internal and inherent to the project, such as those related to failing to prove the steam resource. But in order to mitigate against external risks, such as the inability of the offtaker to meet the terms of the power purchase agreement or nationalisation of private assets, they often seek guarantees from the national and county governments that their investments will be protected. Meanwhile, sustainable charcoal production offers an opportunity to better clarify and manage the distribution of financial risks compared to the dominant informal and unsustainable charcoal trade, where traders, distributors, and government officials pursue their own private economic interests at the expense of the public interest. Promoting sustainability in the charcoal sector can support income generation for the producers and generate tax revenue for the government, which can be reinvested in reforestation programmes or other livelihood initiatives.

The social and environmental risks of geothermal power development and sustainable charcoal production are largely borne by local communities, which are rights holders of the land where the activities generally occur. Kenya's strong land rights mean that traditional communities even maintain some access rights on privately owned lands. Since the livelihood of many of these communities is so closely tied to the land, the national and county governments – as duty bearers for the citizens they govern – have a responsibility to uphold these rights. This is particularly the case where the capacity of rights holders to manage social and environmental impacts is limited. For geothermal, this means effectively regulating private project developers and state-owned electricity utilities. For charcoal, this means ensuring charcoal producer associations maintain established standards and that forest resources are sustainably managed. Environmental and social impact assessments, with associated resettlement action plans, are the typical tools used by duty bearers to hold private economic interests accountable for minimising risks to local communities. But there is potentially a greater role for monitoring by local citizens who might be better placed to identify changes in their local communities and environment.

Weighing risks and benefits

The distribution of risks among powerful and marginalised stakeholders affects how these risks are weighed against benefits during decision-making processes around certain actions (or inaction). In the geothermal sector, the financial benefits accruing to private companies or government from the sale of electricity, along with the political benefit of ensuring reliable power from a clean energy source for the middle and elite classes, appear to have much more weight than the concern over the livelihoods of a few local communities with little influence. But with devolution and the prospect of greater benefit sharing, more weight may be given to those local concerns by the county government.

In the charcoal sector, meeting the growing household energy demand of an increasing urban and peri-urban population appears to hold more weight than concerns over degradation and deforestation in distant locations as a result of unregulated production of charcoal. In the past, more weight has been given to financing demand-side measures to reducing charcoal consumption, such as adoption of efficient cooking technologies. Supply-side measures that potentially have more impact on forest cover and GHG emissions are only recently gaining political attention.

The ways in which risks and benefits are weighed in decisions around geothermal power development and sustainable charcoal production and trade differ depending on the perspective of the stakeholder. And clearly the risk/benefit perspective of those stakeholders who are both more removed from the local landscape, where geothermal power generation and charcoal production activities occur, and more closely connected to where profits from these activities accrue have more influence in decision-making processes. Addressing these spatial and equity dimensions should be of serious concern for those seeking to ensure that Kenya's low-carbon, climate-resilient development pathways benefit all citizens of the country. More transparent and participatory dialogue between private economic interests, duty bearers and rights holders is one way in which the balance could be shifted as Kenya pursues its future development pathways.

Conclusions

In this chapter, we explored the implementation and consequential risks associated with scaling up geothermal power development and sustainable charcoal production, both of which are widely considered core elements of Kenya's low-carbon and climate-resilient development ambitions.

Our research shows that optimism around the potential for greater geothermal power development needs to be tempered with serious action to mitigate against potential social and political risks. Geothermal development in Kenya has largely focused on nurturing a new industry and building technical expertise. But as the sector has grown, so too have the challenges it faces, placing increased pressure on both the government and the private sector to pursue further development in a responsible manner, to ensure the benefits of geothermal development are shared equitably. In the sustainable charcoal sector, barriers to greater adoption of efficient and sustainable forest management, charcoal production, and charcoal consumption practices require co-ordinated efforts to strengthen the capacity of the implementing entities and charcoal producer associations, and to ensure that the enforcing agencies speak to each other in order to address any concerns that may be raised by the market chain actors. In the long term, other cleaner and more sustainable fuels may replace charcoal but, in the short and medium term, investments in the sustainability of this important urban fuel are imperative to ensure that Kenya's natural forest resources are responsibly managed.

Our analysis of implementation and consequential risks associated with pursuit of greater geothermal power development and more widespread adoption of sustainable charcoal production and trade identified a clear distribution of risks across spatial and equity dimensions. The risks and benefits accruing to those marginalised stakeholders located close to natural resource landscapes, and who were often the poorest of all stakeholders, tended to be given less weight than the risks and benefits accrued by those in positions of relative economic and political power. Indeed, our analysis could benefit from action research following how risks and benefits are really weighed by different stakeholders and how that affects decision making in practice. We also advocate further research on the political and social dimensions of low-carbon and climate-resilient energy pathways in Kenya to better understand how these pathways might be realised in an equitable manner.

Notes

1 See https://data.worldbank.org.
2 $MtCO_2e$ = million tons of carbon dioxide equivalent.
3 See www.mtp3.go.ke/.
4 For example, a number of plans under Vision 2030 – such as electrification of the new Mombasa–Nairobi railroad, establishment of the hi-tech Konza City, and development of industrial parks close to geothermal sites – have so far failed to materialise.
5 See equator-principles.com.
6 See https://data.worldbank.org.

References

Bailis, R. (2009). Modeling climate change mitigation from alternative methods of charcoal production in Kenya. *Biomass Bioenergy* 33, 1491–1502. https://doi.org/10.1016/j.biombioe.2009.07.001.

Bailis, R., Wang, Y., Drigo, R., Ghilardi, A., Masera, O. (2017). Getting the numbers right: revisiting woodfuel sustainability in the developing world. *Environ. Res. Lett.* 12, 115002. https://doi.org/10.1088/1748-9326/aa83ed.

Bergek, A., Jacobsson, S., Carlsson, B., Lindmark, S., Rickne, A. (2008). Analyzing the functional dynamics of technological innovation systems: A scheme of analysis. *Research Policy* 37(3), 407–429.

Dalberg Advisors (2018). *Cleaning up cooking in urban Kenya with LPG and bio-ethanol.* SouthSouthNorth, Cape Town.

Dalla Longa, F., van der Zwaan, B. (2017). Do Kenya's climate change mitigation ambitions necessitate large-scale renewable energy deployment and dedicated low-carbon energy policy? *Renew. Energy* 113, 1559–1568. https://doi.org/10.1016/j.renene.2017.06.026.

ESMAP (2012). *Handbook on planning and financing geothermal power generation.* World Bank, Washington, D.C.

Falzon, J., Veum, K., van Tilburg, X., Halstead, M., Bristow, S., Owino, T., Murphy, D., Stiebert, S., Olum, P. (2014). *NAMA for accelerated geothermal electricity development in Kenya: Supporting Kenya's 5000+MW in 40 months initiative and Vision 2030 (No. Proposal).* ECN; Government of Kenya, Nairobi, Kenya.

GEOCOM (2015). *Cost and Financial Risks of Geothermal Projects.* GEOCOM E-Learn. Course. Available at: http://geothermalcommunities.eu/elearning.

Government of Kenya (2007). *Kenya Vision 2030.*

Government of Kenya (2009). *Forest (Charcoal) Rules, 2009.*

Government of Kenya (2010a). *National Climate Change Response Strategy.*

Government of Kenya (2010b). *The Constitution of Kenya 2010.*

Government of Kenya (2011). *Least Cost Power Development Plan 2011–2030.*

Government of Kenya (2013). *National Climate Change Action Plan 2013–2017.*

Government of Kenya (2016a). *Kenya National Adaptation Plan 2015–2030.*

Government of Kenya (2016b). *Climate Change Act.*

Government of Kenya (2017). *The Energy Bill, 2017.*

Government of Kenya (2018a). *National Climate Change Action Plan 2018–2022.*

Government of Kenya (2018b). *Least Cost Power Development Plan 2017–2037.*

Hekkert, M.P., Suurs, R.A.A., Negro, S.O., Kuhlmann, S., Smits, R.E.H.M. (2007). Functions of innovation systems: A new approach for analysing technological change. *Technological Forecasting & Social Change* 74, 413–432.

Hosonuma, N., Herold, M., De Sy, V., De Fries, R.S., Brockhaus, M., Verchot, L., Angelsen, A., Romijn, E. (2012). An assessment of deforestation and forest degradation drivers in developing countries. *Environ. Res. Lett.* 7, 044009. https://doi.org/10.1088/1748-9326/7/4/044009.

Johnson, O., Nyambane, A., Cyoy, E., Oito, L.G., 2016. *County energy planning in Kenya: Local participation and local solutions in Migori County (No. SEI Working Paper 2016–01).* Stockholm Environment Institute (SEI), Nairobi.

Kamfor Company (2002). *Study on Kenya's energy demand, supply and policy strategy for households, small scale industries and service establishments (Final Report).* Ministry of Energy, Nairobi, Kenya.

Kapika, J., Eberhard, A. (2013). *Power-sector reform and regulation in Africa: lessons from Kenya, Tanzania, Uganda, Zambia, Namibia and Ghana.* HSRC Press, Cape Town.

Karekezi, S., Mutiso, D. (2000). Power Sector Reform: A Kenyan Case Study, in: *Power Sector Reform in Sub-Saharan Africa.* Palgrave Macmillan, London.

Karekezi, S., Turyareeba, P. (1995). Woodstove dissemination in Eastern Africa – a review. *Energy Sustain. Dev.* 1, 12–19. https://doi.org/10.1016/S0973-0826(08)60094-0.

Kenya Power (2018). *Annual Report and Financial Statements for the Year Ended 30 June 2017.* Kenya Power, Nairobi.

Kiruki, H.M., van der Zanden, E.H., Malek, Ž., Verburg, P.H. (2017). Land Cover Change and Woodland Degradation in a Charcoal Producing Semi-Arid Area in Kenya: Charcoal, Land Cover Change and Woodland Degradation in Kenya. *Land Degrad. Dev.* 28, 472–481. https://doi.org/10.1002/ldr.2545.

Kojima, M., Bacon, R., Zhou, X. (2011). *Who Uses Bottled Gas? Evidence from Households in Developing Countries*. Policy Research Working Papers. The World Bank. https://doi.org/10.1596/1813-9450-5731.

Kubo, B.M. (2003). *Environmental management at Olkaria geothermal project, Kenya*. Presented at the International Geothermal Conference, Reykjavík, Iceland, pp. 72–80.

Mariita, N.O. (2002). The impact of large-scale renewable energy development on the poor: environmental and socio-economic impact of a geothermal power plant on a poor rural community in Kenya. *Energy Policy* 30, 1119–1128.

Matara, E., Sayagie, G. (2018). Let counties benefit from their assets. *Daily Nation*.

Ministry of Energy and Petroleum. (2013). *5000+MW by 2016: Power to Transform Kenya (Investment Prospectus 2013–2016)*. Republic of Kenya.

Ministry of Energy and Petroleum (2016). *Sustainable Energy for All (SE4ALL): Kenya Action Agenda*.

Ministry of Environment and Natural Resources (2013). *National Environment Policy*.

Ministry of Environment and Natural Resources (2015). *Kenya's Intended National Determined Contribution (INDC)*.

Ministry of Environment and Natural Resources (2016). *Kenya Strategy Investment Framework for Sustainable Land Management 2017–2027*.

Ministry of Environment, Water and Natural Resources (2013). *Analysis of the charcoal value chain in Kenya*.

Musembi, R. (2014). *GDC's geothermal development strategy for Kenya: progress and opportunities*. Presented at the Power Africa–Africa Union Commission Geothermal Roadshow, Portland, Oregon, October 2014.

Mutimba, S., Barasa, M. (2005). *National charcoal survey: exploring the potential for a sustainable charcoal industry in Kenya*. Energy for Sustainable Development Africa (ESDA), Nairobi, Kenya.

Mwampamba, T.H., Ghilardi, A., Sander, K., Chaix, K.J. (2013). Dispelling common misconceptions to improve attitudes and policy outlook on charcoal in developing countries. *Energy Sustain. Dev.* 17, 75–85. https://doi.org/10.1016/j.esd.2013.01.001.

Mwangi, M. (2005). *Country update report for Kenya 2000–2005*. Presented at the World Geothermal Congress 2005, Antalya, Turkey, pp. 24–29.

Ndegwa, G., Anhuf, D., Nehren, U., Ghilardi, A., Iiyama, M. (2016a). Charcoal contribution to wealth accumulation at different scales of production among the rural population of Mutomo District in Kenya. *Energy Sustain. Dev.* 33, 167–175. https://doi.org/10.1016/j.esd.2016.05.002.

Ndegwa, G., Nehren, U., Grüninger, F., Iiyama, M., Anhuf, D. (2016b). Charcoal production through selective logging leads to degradation of dry woodlands: a case study from Mutomo District, Kenya. *J. Arid Land* 8, 618–631. https://doi.org/10.1007/s40333-016-0124-6.

Ng'ang'a, J.N. (2005). *Financing geothermal projects in Kenya*. Presented at the Workshop for Decision Makers on Geothermal Projects and Management, UNU-GTP and KenGen, Naivasha, Kenya, p. 10.

Ngugi, P.K. (2012). *Kenya's plans for geothermal development – a giant step forward for geothermal*. Presented at the Short Course on Geothermal Development and Geothermal Wells, UNU – GTP and LaGeo, Santa Tecla, El Salvador, p. 8.

Njenga, M., Karanja, N., Munster, C., Iiyama, M., Neufeldt, H., Kithinji, J., Jamnadass, R. (2013). Charcoal production and strategies to enhance its sustainability in Kenya. *Dev. Pract.* 23, 359–371. https://doi.org/10.1080/09614524.2013.780529.

Ogola, P.F.A., Davidsdottir, B., Fridleifsson, I.B. (2012). Potential contribution of geothermal energy to climate change adaptation: A case study of the arid and semi-arid eastern Baringo lowlands, Kenya. *Renew. Sustain. Energy Rev.* 16, 4222–4246.

O'Keefe, P., Raskin, P., Bernow, S. (Eds.). (1984). *Energy and development in Kenya: opportunities and constraints.* Beijer Institute, Royal Swedish Academy of Sciences; Scandinavian Institute of African Studies, Stockholm and Uppsala, Sweden.

Ole Koissaba, B.R. (2018). *Geothermal Energy and Indigenous Communities: The Olkaria Projects in Kenya.* Heinrich Böll Foundation, Nairobi, Kenya.

Omenda, P., Simiyu, S. (2015). Country Update Report for Kenya 2010–2015. *Proc. World Geotherm. Congr. 2015* 11.

Riaroh, D., Okoth, W. (1994). The geothermal fields of the Kenya rift. *Tectonophysics* 236, 117–130. https://doi.org/10.1016/0040-1951(94)90172-4.

Shortall, R., Davidsdottir, B., Axelsson, G. (2015). A sustainability assessment framework for geothermal energy projects: Development in Iceland, New Zealand and Kenya. *Renew. Sustain. Energy Rev.* 50, 372–407. https://doi.org/10.1016/j.rser.2015.04.175.

Simiyu, S.M. (2008). *Status of geothermal exploration in Kenya and future plans for its development.* Presented at the Geothermal Training Programme 30th Anniversary Workshop, Reykjavík, Iceland, p. 10.

Tigabu, A. (2017). Factors associated with sustained use of improved solid fuel cookstoves: A case study from Kenya. *Energy Sustain. Dev.* 41, 81–87. https://doi.org/10.1016/j.esd.2017.08.008.

Wanjiru, H., Nyambane, A., Omedo, G. (2016). *How Kenya can transform the charcoal sector and create new opportunities for low-carbon rural development.* Stockholm Environment Institute, Nairobi, Kenya.

Wanleys Consultancy Services (2013). *Analysis of Demand and Supply of Wood Products in Kenya.* Ministry of Environment, Water and Natural Resources, Nairobi, Kenya.

Zulu, L.C., Richardson, R.B. (2013). Charcoal, livelihoods, and poverty reduction: Evidence from sub-Saharan Africa. *Energy Sustain. Dev.* 17, 127–137. https://doi.org/10.1016/j.esd.2012.07.007.

Part VI

Synthesis

14 Transition pathways, risks, and uncertainties

*Jenny Lieu, Susanne Hanger-Kopp,
Wytze van der Gaast, Richard Taylor, and
Ed Dearnley*[1]

Introduction

Transition pathways to reduce carbon emissions and other greenhouse gases (GHGs) in energy-related sectors are often portrayed as inherently positive changes to support climate mitigation efforts. The most significant concerns around them are generally whether we successfully implement the necessary technology, policies, and behavioural changes needed to follow the pathway. However, transitions may also involve potentially negative impacts on the economy, society, and environment, for example in the form of hidden costs, unemployment, or the loss of biodiversity. This implies two dimensions of risk, as we discuss in Chapter 2 and employ across the case studies presented: The first is the risks associated with designing or implementing the pathway, and the second is the unintended yet potentially negative consequences resulting from following a pathway.

We used narratives as an enriched form of insight to provide comprehensive knowledge from a wider systematic research and engagement process carried out over three years of the TRANSrisk research project. These narratives from Europe, Asia, Africa, and North and South America discuss a diverse set of pathways that may contribute to low-carbon transitions. In some cases the focus is on the past and how lessons learned may influence future transition efforts (e.g. Spain, Greece). In others, pathways were designed as high-level strategies through a top-down approach but require additional bottom-up knowledge and consideration to be successfully implemented (e.g. China, Switzerland). Finally, in some cases pathways were created based on stakeholders' perceptions and preferences, or co-created with stakeholders reflecting desired futures for a certain technological innovation or sector (e.g. Austria, Indonesia, Canada).

We synthesise our findings across narratives in three nested dimensions present in all pathways: *technological innovations* implemented in an existing or new market; *policy mixes* supporting these technologies; and *society* where impacts are felt, perceptions created, and support or opposition is determined. All three dimensions are interrelated, but as conceptual boundaries they provide a starting point to explore risks identified as barriers to transitions and potential negative consequences of a pathway.

Technological innovations: a starting point towards low-carbon transitions

Each case begins with one or more technological innovations that have the potential to support a low-carbon transition. They roughly can be summarised under four major technology-related themes: (1) technologies to reduce emissions for incumbent, large-scale sectors that are dependent on non-renewable resources; (2) renewable energy technologies to support a country's mainstream electricity generation mix; (3) energy efficiency technologies to reduce emissions, primarily in the building sector; and (4) technologies linked to livelihood activities and the corresponding supply chains (Table 14.1).

Financial barriers to deploying technological innovations

Across the board, the narratives highlight how innovation not only needs to be technically feasible but also financially feasible. The majority of stakeholders identified high upfront costs and unstable market situations (compared to those for fossil-based alternatives) as major barriers to the investment needed for both incumbent large-scale centralised technologies and new decentralised, small-scale technologies. More specifically, investors consider investment in renewable technologies to be a higher risk compared to incumbent technologies, which increases the cost of borrowing capital. Investors willing to take on this risk require reassurance, for example by revenue guarantees. This was highlighted for both the diverse cases of geothermal in Kenya and new nuclear power in the UK. Large-scale solar projects in the Netherlands face similar investment challenges, and stakeholders additionally highlight the costs for changes and upgrades to the existing electricity infrastructure. This may seriously affect the financing capacity of both public and private stakeholders, who might look for lower opportunity costs and higher revenues elsewhere. Private investment is also crucial for energy efficiency measures in the building sector, yet in some cases is not readily available. In China only some city authorities provide funding for energy efficiency measures, while in Greece there is recognition and will to push forward energy efficiency technologies in buildings but a lack of public funds.

Small-scale technologies that have a direct interface with end users can encounter cost barriers at the household level, where the upfront cost of these technologies is born. Programmes to overcome cost barriers are often regionally specific and have a limited timeframe. This fails to overcome challenges in scaling up deployment of the technology, which is required to make a more substantial impact on emissions. In Bali, the cost of biogas systems was identified as a barrier as most farmers cannot afford biogas without a government subsidy. While cost is not a key barrier for solar rooftop installations in the Netherlands, households critically consider payback times and rate of return when comparing competing investments.

Table 14.1 List of country studies and technological innovations to support low carbon pathways

Country study	Technology sector
Pathways for incumbent large-scale technology systems	
Austria	Potential to switch to a hydrogen-based energy system for the iron and steel sector
Canada, Alberta (Athabasca oil sands)	Promoting technological innovations to reduce emissions that include carbon capture and storage and reductions in methane flaring
United Kingdom	Considers nuclear power expansion with potential for small modular reactors
Pathways towards renewable electricity systems	
Chile	Explores a mix of renewable energy including solar power and hydropower
The Netherlands	Potential to scale up solar PV in the built environment and for larger scale solar PV in parks
Spain	Evaluates the roll out of renewable energy and highlights the role of solar
Switzerland	Options to scale up renewable energy for solar, wind, and geothermal
Kenya	Considers an expansion of renewables through geothermal energy
Pathways energy efficient building sectors	
China	Considers incorporating energy efficiency technologies within its built environment in the residential building sector
Greece	Explores energy efficiency technologies in both public and private (residential and service) buildings
Pathways using renewable household-level technology at the community level	
Indonesia (Bali)	Investigates the potential for biogas applications in its agriculture sector for cooking
Kenya	Explores efficient retort kilns and cook stove improvements based on sustainable charcoal technologies

Unintended environmental impacts of low-carbon technologies

Many stakeholders expect low-carbon technologies to lead to positive environmental impacts compared to the status quo. These most obviously occur through climate change mitigation but also through national and regional environmental benefits. However, as various stakeholder groups at the local level have explained, when implemented and/or scaled up these technologies can also lead to consequential risks with environmental impacts. These are a consequence of a variety of factors but most often are linked to land-use changes and other knock-on impacts in the value chain of technology production, installation, operation, and dismantling.

Land-use change and a varying set of associated impacts on natural habitats, migratory routes, and environmental pollution can occur from fossil-based energy production, such as negative impacts for the oil sands in Alberta. Stakeholders also highlight negative impacts with the expansion of renewables, including geothermal power in Kenya and renewables in the UK.

In Alberta, the development of the oil sands, even with reduced emissions, will continue to affect the environment. Locally, these impacts result from land-use changes in the oil sands extraction processes, which causes damage to ecosystems. Additionally, oil sand waste products pollute the air and local rivers. Land-use changes can also occur with the development of renewable energy, as seen in the UK renewable energy sector and Kenyan geothermal development. In the UK, stakeholders highlighted negative impacts on nature preservation with the expansion of renewable energy installations, rather than environmental concerns linked to nuclear accidents and waste. In Kenya, geothermal power facilities were the cause of unsustainable water abstraction from lakes and rivers, as well as disturbances to natural habitats and wildlife migratory routes.

In Bali, biogas installations, which turn animal waste into household biogas for cooking and lighting, reduce CO_2 emissions but also emit methane if the biogas is not burned. Methane has a much higher global warming potential per unit emitted than carbon dioxide. The efficient use, rather than an over or underuse, of cookstoves technologies is necessary to ensure a sustainable transition. In Austria, hydrogen-based steel technology, supported by a clean electricity mix, is anticipated to be a crucial part of the low-carbon energy transition in the iron and steel sector. As part of decarbonising the electricity mix, energy storage solutions need to be scaled up, which could lead to negative environmental impacts due to the extraction and disposal of materials including lithium for batteries. Additionally, the disposal of renewable energy systems also needs to be considered, for example wind turbine rotors built with composite materials that are difficult to recycle. These impacts can be expected in most cases where renewables will be scaled up but are rarely considered by the wider stakeholder community.

Policy mixes to support low-carbon technological innovations

Technological innovations rely on public and/or private policies to gain traction for eventual upscaling. All of our narratives discuss existing and/or new policies that would support decarbonisation technologies at different scales. We observed some trends on the discussion of risk in policy mixes across the country studies. We will draw from specific examples to highlight policy-related risks in different pathways. Across countries, policies aim to create a favourable climate for investment supporting implementation of individual technology projects or scaling up deployment. During the implementation of policies, monitoring and enforcement are identified as important elements that could create or enable barriers against supporting the innovation. Additionally, policies and politics are often intertwined. If a technological innovation is essential to the current political agenda, policies are often advanced to support it.

Policy instruments to encourage private investments for technological deployment

We highlighted that most stakeholders emphasised the need for substantial public and/or private investments. Investors are known to be keen on stable markets with reliable revenue streams, while renewable energy options are characterised as unstable or unreliable in terms of revenue. Policy instruments and initiatives backed by government provide an indication for the longer-term market potential of a specific technological option; for example, governmental guarantees about project-level revenue streams can reduce financial implementation risks.

Policy instruments discussed in our case studies include economic incentives to support low-carbon technologies such as: subsidies for renewable energy (the Netherlands, Indonesia); feed-in tariffs (FITs) for renewable electricity generation (Spain, Greece, Switzerland) or other price guarantees (UK nuclear power) backed by government-endorsed policies. There are also policy disincentives such as carbon or energy taxes (Chile and Canadian carbon tax), which are used as a means of penalising high-carbon emitters. While economic policies are widely applied across case study countries, there are also non-economic policy instruments that can drive forward a transition. These include high-level energy plans, for example Kenya's national and sectoral acts and plans and EU renewable energy targets impacting the Austrian steel sector. They also include standards and performance requirements for individual technologies or sectors (building codes and standards for energy efficiency in China and Greece).

Across our case studies, stakeholders identified a set of risks in the implementation of different policy instruments. While the aim of policies was to reduce financial barriers, the policies themselves can lead to negative consequences, which illustrates the need for selecting and fine-tuning policy instruments for preventing and overcoming barriers. There are still uncertainties in how policies can effectively support the scaling up of renewable technologies in different country contexts. As can be observed from previous trends, such as the renewable energy sector in Spain, economic incentives such as feed-in tariffs and premium tariffs can significantly boost renewable energy generation. However, in Spain the overheated market significantly increased regulatory costs and retroactive measures were put in place on renewables to cut existing subsidies. The absence of continued policy support for renewable energy caused a market collapse, an unanticipated consequential risk in the Spanish renewable energy sector. Hence, the case provided a clear example of the need for carefully selecting instruments for effective policymaking and coherent policy mixes.

An example of a subsidy policy is seen in the Dutch sustainable energy subsidy scheme to support the adoption of, among other options, large-scale ground-mounted solar power projects. Investors receive a government guarantee of a stable revenue stream for a period of 15 years, which addresses the main implementation risk of this technology option. While the scheme reduces risks for solar park investors, it creates an extra financial burden and uncertainty for the government around the future costs of expanding sustainable energy. The Dutch government mitigates this risk by maximising the annual subsidy amounts under the scheme. However, for the longer term, it is uncertain how large the subsidy schemes will become as this will depend on cost development of solar PV systems, electricity prices, and European energy and climate directives and how these are incorporated into national energy and climate policymaking.

In Indonesia, national and regional subsidy programmes partially or fully subsidise individual biogas digesters in order to reduce the barriers of initial capital investments for biogas digesters. However, policies do not explicitly support larger-scale biogas-for-electricity generation and renewable electricity generation. Many of the issues are linked to subsidies for coal-powered generation. Even then, fully subsidised programmes have not necessarily lead to the successful uptake of the technology. In fact, biogas programmes that were highly subsidised were often less successful than lightly subsidised schemes, due to limited personal investment and a low sense of ownership among the farmers involved. If leakage in the tank occurred during biogas operation, the farmer would often abandon the installation due to high cost of repair and lack of support. Thus, ongoing maintenance or warranty, supported by a monitoring programme, is also needed in addition to subsidies. This will be discussed further in the next section.

Incoherence across the design, implementation, monitoring, and enforcement of policies

In most cases presented in this book, stakeholders see implementation risks through the particular lens of the policy implementation process. We observed a lack of coherent policy design and implementation process (including monitoring and enforcement) across many country studies. The design, implementation, and monitoring of policies is influenced by policymakers and their corresponding institutions, along with politicians and the governments who support the policy goals and technologies behind the low-carbon transitions. Challenges occur when policies are implemented, usually requiring co-ordination across various government institutions at different governance levels to ensure effective monitoring and enforcement. Finally, there may be barriers to implementing these policies due to poor monitoring and enforcement procedures within the existing governance structure. We see examples of these across different case studies.

In Austria there is no clear strategy to promote hydrogen-based technologies, which would be important for addressing energy storage and reliable supply issues. The absence of an overarching target and concrete implementation strategy has led to uncertainties in their low-carbon transitions. There is also a lack of agreement across local, regional, and national-level institutions, which all have different interests and prevents the co-ordination of policies. Likewise, in Switzerland, cantons and federal governing bodies interpret the national energy strategy differently, specifically with respect to objectives for energy efficiency, energy supply, support for investments, and environmental protection. The challenge in the Swiss context is that there is no common space to harmonise the differences across governance levels, creating risks in the policy implementation process.

In Indonesia there have been challenges with implementing biogas programmes due to the conflicting priorities and motivations at the national and sub-national governmental levels. Consequently, there was no streamlined approach to monitor the progress of biogas programmes across different government institutions and implementing agencies. Monitoring is essential for assessing the outcomes of biogas installations and to identify barriers including technical difficulties in the operations of the digester. As a result of this lack of monitoring and support many biogas installations were abandoned, a negative consequence of the renewable energy programme.

The charcoal sector in Kenya also suffers from weak enforcement of the formal permitting system set up to promote the sustainable production, transport, and use of charcoal. Most charcoal circulates in the informal sector and corruption is widespread. Transporters and wholesalers are compelled to pay bribes to traffic police, who do not fully understand the formal permitting systems. Some Kenyan counties have set their own regulations with local charcoal producer associations and transporter associations, and there are still uncertainties around whether local initiatives are more effective than national initiatives.

Energy efficiency certification policies in China's building sector have been less successful than anticipated, even though the growth of green buildings has been fuelled by policies and targets set by the Chinese government. One of the main challenges identified by government stakeholders is the immature green certification implementation process, along with weak monitoring and enforcement of policies in the operation of new buildings. Additionally, industry players in the construction market are not taking a lead as there are no policy incentives to encourage their involvement with the certification scheme.

The Spanish case explicitly highlights how a low-carbon energy transition is by no means a linear process but is characterised by both failures and successes. The particular design of the chapter highlights the need for policy learning, which is an inherent part of the policy process and implicitly important for all cases.

Influence of political will on technology selection and policy promotion

Politicians are more hesitant to support low-carbon pathways if they imply a risk of disrupting the socio-economic balance within their countries. The (lack of) support for technologies can be primarily driven by political will, creating a strong enabler for favoured technologies but also high risks if political backing is withdrawn. We see examples throughout the country studies where policies and political support create barriers for policy implementation. This directly relates to carbon lock-in effects, i.e. barriers to new technologies, as established technologies have become tightly entangled with the existing institutional set-up (e.g. Unruh, 2000).

As these disruptions are often noticeable in the short term and directly affect voters, policies supporting low-carbon pathways can be subject to change due to elections. An example of risks caused by political change can be observed in the oil sands sector in Canada. A new political party (the NDP) came into power for the first time in 2015, uprooting the incumbent party who were large supporters of oil sands development. The NDP implemented a carbon tax for oil sands production which has impacted profits and has been unpopular within the industry. If a different political party gains power through the next election in 2019 they may reverse the carbon tax. In Indonesia, where fossil-fuel exports are an important source of government revenue and economic development, changes in political leadership can lead to significant uncertainties for the biogas renewable energy programmes as there is no certainty that the new government will continue to support programmes.

Without financial support some low-carbon technologies, particularly those needing intensive capital investment, would not be feasible. New nuclear power stations in the UK, for instance, face a high risk in securing funding. The high costs of construction and nuclear waste handling requires secure long-term revenue streams to ensure financial viability. For that, there has been significant political support for the development of the new Hinkley Point C nuclear power plant. The government has negotiated with investors to secure long-term fixed electricity price contracts that are well above current wholesale prices. This example shows how, through a long-term guarantee scheme, the impacts of changes in the political context can be kept relatively small.

In Kenya, on the other hand, the lack of strong, long-term guaranteed political support has led to implementation risks for geothermal plants. One of the barriers for these plants is securing funding. Since geothermal plants are viewed as higher risk investments, investors need the reassurance of financial and political stability for their investments. A means of reducing the financing risk for the project developers is to guarantee the revenue stream through a power purchase agreement with Kenya Power, which is jointly owned by the government and private investors. But there have been delays in finalising this financing as the government has been slow to issue letters of support for the projects. Due to this delay, investors face a higher risk that funding for the projects will not be secured.

In Greece, the overarching policy strategy for energy efficiency is in place but there is a lack of detailed economic incentives, subsidies, and tax exemptions. These policy instruments depend on political will and the government's financial capacity. The economic recession and other pressing socio-economic issues have impacted the government's financial capacity to push forward energy efficiency initiatives and provide longer-term stability and support. Political instability in recent years, including snap elections and the inability to form single-party governments, has side-lined climate change policy and action on the political agenda due to more pressing socio-economic challenges.

Policy mixes and their implementation, monitoring, and enforcement processes require careful consideration in a low-carbon transition pathway. While policies may initially help to overcome financial barriers, they may not be successful in the longer run due to poor implementation, weak monitoring, and poor enforcement. On the other hand, even if all of these policies and policy processes are set in place, politics may unravel the carefully chosen policy support package for a low-carbon pathway. Politics can be unpredictable and is itself an implementation risk, thus additional support for low-carbon technologies may be needed from societal stakeholders who may (in)directly influence politics. While the societal dimension can be an enabler in a pathway, it can also imply risks, such as resistance to low-emission technology options. This in itself adds another level of complexity as discussed in the next section.

Societal impact on technologies: technologies' impact on society

A third critical determinant of the success of a pathway is whether it meets and resonates with people's needs and priorities. While a technological and economic focus can indicate the technical potential and economic affordability of a low-carbon pathway, understanding societal priorities and concerns is a key determinant of the realistic potential for expanding a low-carbon technology. The narratives in this book show several examples of this, including negative impacts, for example concerns about job losses and resistance to solar parks, to more positive support for technologies as an enabler for people to personally contribute to addressing the climate problem.

The examples of societal impact on the deployment of low-carbon technologies vary from individual household behaviours to wider societal acceptance at the (sub)national level. Additionally, the impact of technologies and policies on people has been frequently highlighted as both a barrier and consequential risk for transition pathways. Policies intended to support low-carbon technologies often have positive environmental impact; yet policies can also lead to societal injustice if the higher cost of the technologies are borne by the most disadvantaged in society.

The influence of socio-economic interests on technological selection

The socio-economic interest to maintain the steel, oil sands, and nuclear power sectors in Austria, Canada, and the UK demonstrate how the historical development of sectors can be entrenched within society. These sectors become intertwined with the development of a low-carbon pathway despite the risks associated with each technology. As a consequence, when a country has a strongly entrenched socio-economic context, the most feasible way forward can be improvement of existing technologies: even when they are unattractive from an environmental (including climate) perspective they can nevertheless be more realistic from a social perspective.

There is a long-standing tradition of high-quality steel production in Austria. The sector has close interlinkages with other sectors of the economy, so that low-carbon measures in steel production may have social implication for other sectors. The steel sector has long been important for jobs and prosperity and a transition to a lower-carbon option may cause job losses. However, the sector also acknowledges that technological innovation, as part of a shift to a low-carbon pathway, is required if the industry is to survive and thus secure employment in the longer term. A similar situation is seen in Alberta, Canada, where the Athabasca oil sands are located. The development of the oil production industry is closely linked to the province's social and cultural identity, as well as its long-term economic prospects.

Viewed through the global lens of Paris Agreement targets, supporting the development of both incumbent sectors in Austria and Canada may prevent a transition away from carbon-intensive economies as, while their carbon intensity can be reduced, reaching zero carbon emissions in these sectors is highly unlikely. Nevertheless, within the (sub)national context, both sectors will continue to be supported socially and thus politically for the foreseeable future. Reducing emissions in the Austrian steel sector and the Albertan oil sands are viewed as feasible transitions for the respective local stakeholders; eliminating these high-carbon sectors altogether is not desirable given the existing social context, and thus not realistic to consider. Likewise, the UK nuclear power sector is rooted in a history of nuclear technology for both military and energy production stretching back to the 1930s. Nuclear energy currently enjoys significant political support in the UK and is seen as part of the transition to a lower-carbon economy, despite the risks associated with nuclear waste and the potential for accidents.

Societal acceptance as a driver to support low-carbon technologies

Although the risks associated with scaling up renewable energy technologies are known, several countries have chosen this pathway to reduce emissions. One example is Switzerland, which previously made large investments in nuclear power. The Swiss have opted to close down nuclear power plants due to the perceived risks of nuclear accidents, as observed by the Fukushima Daiichi nuclear accident of 2011. The conscious decision to move away from nuclear power required societal consensus in Switzerland. Such a consensus is easier to achieve when alternatives technologies are available that fit well within the country's landscape and do not have significant negative impacts on people's livelihoods. In the Swiss case, social acceptance for expanding renewables was relatively easy to achieve as alternative resources (such as hydropower) are available which meet with little social resistance. Social acceptance is particularly important for renewable energy technologies as these tend to be more decentralised and are usually geographically scattered (e.g. wind power farms and solar PV in Spain, Netherlands, and Chile). If a technology or action requires individuals or households to respond, as seen in solar roof top installations (Spain, Netherlands, Chile), stakeholders' perspective becomes critical for the success of implementing the pathway.

Large, centralised renewable energy technologies tend to be constructed on green field sites. These require government and political support, and often also the approval from impacted communities through the planning system. In the Netherlands, larger ground-mounted solar power parks are being promoted by policy incentives (as discussed earlier) but, unlike solar rooftops on existing buildings, larger solar parks will need to carefully consider spatial planning requirements. Project developers proposing to install solar parks near communities need to consult with community stakeholders as early as possible to avoid the risk of public opposition. Opposition can delay or halt project implementation, as seen in some projects in the northern regions of the country. Consulting with community stakeholders impacted by a proposed solar park allows them to be part of the project process and provides opportunities for stakeholders to suggest modifications to benefit from the project. The Dutch narrative contains examples of how engaging stakeholders as early as possible in the decision-making process can reduce risks and act as a key enabler.

Land use linked to energy projects can have both a negative environmental impact and also adverse consequences for local communities. These communities may protest against large energy projects, as seen with traditional communities in Kenya. In Kenya, the livelihoods of traditional communities are closely linked to the land. The county and national government are therefore responsible for upholding these rights. Land rights are strongly respected and traditional communities can continue to access privately held land. In Kenya, environmental and social impact assessments are required of industry players, including geothermal developers and operators in the charcoal supply chain. They are responsible for minimising the risks to the impacted communities, for example through adequate monetary compensation or by making jobs available to affected communities.

Similarly, in Canada, Indigenous communities located near oil sands are adversely impacted due to land-use changes. Aside from health impacts, emissions and waste from these developments alters the landscape and changes the community's ability to exercise their traditional hunting and gathering practices. Energy companies who developed close to traditional sites are facing litigation due to the negative impacts of their development. Failing to consult with Indigenous communities prior to developing in the oil sands project has led to project delays and could even halt projects altogether.

Challenges in end-user behaviour: change in the uptake of technologies

In the narratives described, we have seen that behavioural changes often present a major barrier in pathways that require changes at the household or end-user level. End-user stakeholders often consider the immediate benefit or disadvantage of adopting a particular behaviour (energy efficiency) or technology (solar power, efficient cookstove technology). For instance, a government-funded programme fully paid for biogas installations in Bali, Indonesia, but the technology was not successful due to breakages and a lack of operational support. Moreover, ensuring enough feedstock to generate a sufficient quantity of biogas led to a lower sense of ownership on the farmers' side. Mobilising efforts at the farmer level is challenging; upscaling biogas digesters as a means for village-wide generation was not seen as realistic by some, but there may be more potential for large-scale biogas applications for electricity. End-user behaviour has an important impact on applied technologies, which is evident in efforts to promote efficient cookstoves in Kenya. Efficient cookstoves are part of a wider initiative to support more sustainable charcoal production together with efficient retort-kilns, standards, and certification. Improved efficiencies in cookstoves, however, may inadvertently lead to rebound effects, including more intensive use of cookstoves when compared to previous usage patterns.

Other impacts of end-user behaviour are seen in the case of energy efficiency retrofits in buildings across different climatic zones in China. Energy-efficient measures in the northern regions promote improved heating management, which involves indoor temperature control and net-metering installations. While the scheme was successful in the colder and drier northern regions, energy efficiency programmes have had limited impact in cities with different climates. In the southern regions, such as Shanghai, the climate is more humid and residents keep their windows open throughout the year to manage indoor comfort. Retrofits carried out to improve thermal insulation in Shanghai were offset by residents opening windows and negating the impact of the building's improved thermal performance.

Impacts of low-carbon technologies on social justice

Technologies and policies intended to reduce emissions may unintentionally have a negative impact on social justice, including gender equality. In Chile and Greece there were several examples where policies and technologies led to financial burdens that primarily fell on the economically disadvantaged. In Chile, energy poverty has been identified as a consequential risk for households as a result of the carbon tax implemented to support renewable energy generation. Energy poverty reduces the ability of households to maintain comfortable indoor temperatures. It can have negative impacts on health, particularly with a changing climate that may require more heating or cooling. Heating and cooling needs also vary depending on the different climate zones and energy resources available across Chile. The types of energy resources also impact the cost and extent of the economic burden on households; for example, solar power is costlier that traditional biomass used in the less affluent southern region of the country. Assessment of energy poverty in Chile does not currently consider the different thermal zones and varying household incomes levels. There are also external events that impact energy poverty, including the price of imported natural gas which is beyond the influence of national policies and actions but also increases the risk of fuel poverty in Chile.

In Greece, poverty is viewed by stakeholders as a potential negative consequence of climate and energy efficiency actions. There is a risk that infrastructure and building renovation projects intended to support economically disadvantaged households could inadvertently cause greater hardships. For example, low-interest-rate loans for lower-income households are expected to provide financial support for energy efficiency measures, but the economic recession has made it increasingly difficult for these households to repay loans. On the other hand, these programmes could potentially boost the economy if a longer-term perspective is taken for funding programmes and providing financial support.

Low-carbon initiatives and policies can also result in unequal gender power dynamics as seen in the installation of biogas systems in Balinese farms. Wood fuel used for cooking is often collected by women, who also prepare meals, while the men manage the livestock on the farms. Biogas installations would benefit women by reducing the time require to collect wood, while men will need to increase their time on the farm to collect the organic waste for biogas feedstock. At the household level, male farmers were concerned with the changes in time commitment linked to biogas installations, which policymakers did not view as an issue. However, while the perception is relevant for some parts of Bali, it does not represent the gender-roles across the whole of Indonesia.

In Spain, researchers identified a consequential risk linked to unbalanced gender participation in the energy sector as well as in climate policymaking processes. The energy sector has the lowest proportion of women in executive roles compared to any other economic sector. In climate policymaking there is also a lack of balance, as evident in the 14-strong all-male expert panel selected to provide insights into the design of national energy law in Spain. The lack of a gender balance is also observed in the composition of stakeholder participants responding to the Spanish case study survey. The under representation of women across studies, industry, and policymaking processes can be concerning as decision-making mechanisms supporting the energy transition could fail to address gender-specific issues and needs.

Risk perception of stakeholders across space and time

We consider scale and time to be relevant in all narratives, spanning the dimensions of technology, policy, and society. Most prominently, scale refers to socio-political boundaries, such as different governmental jurisdictions including: city and city-region levels (building sector in the Chinese cities), provincial level (oil sands production in Alberta and biogas in Bali), national level (rolling out solar power in Spain and Chile), and international level (European steel sector impacting Austria). Inconsistencies between sub-national and national priorities were also observed, which do not necessarily align with international priorities. For instance, technological innovations and the supportive policies needed are not necessarily co-ordinated. At the sub-national level (cities/counties/state/provinces), policies have been implemented with more success in some regions than in others, in part due to the power of regional and local governance to enact supportive policy or local by-laws. Additionally, polices at the EU level or national level broadly indicate overarching energy and climate goals, but at the local level implementation can come down to a specific use of space (e.g. land use). The perception of risk at the local level was strongly related to how stakeholders understood and experienced the impact of a technology or policy in their social space including the ecological environment. We see this in the country studies that highlight the importance of different climatic zones and socio-economic realities when implementing policies and technologies (e.g. China and Chile).

Scale is also relevant with respect to the size of investment needed in a specific centralised or decentralised technology. The investment required for different scales of technologies (large- versus small-scale technologies) has implications on the magnitude of the investment risk and the level of reassurance required to secure the investment, or formal government support, needed to overcome the cost risk (e.g. geothermal in Kenya, nuclear power in the UK). Scale is also important when it comes to rolling out technologies across a wider geographical region which, although the logical next step once technologies have been successfully tested, is an often-neglected aspect. This scaling up can complicate both implementation and consequential risks as problems that occur at the project level can be magnified and have wider social-economic and environmental impacts that are more difficult to address and contain (scaling up biogas in Indonesia and solar power in Greece and the Netherlands).

Time is another consideration in the pathways, where the short term (years), medium term (decades), and long term (century) are relevant. Time often intersects with other scales, as already indicated, and influences risk perception and the urgency of actions required to mitigate risks. Adverse climate change impacts are often perceived as a long-term problem from a global perspective, as indicated in the Paris Agreement. In the country studies, most stakeholders perceived their problems within the boundaries of the immediate future (several years) and only some considered a medium time horizon (several decades). Very few stakeholders framed their problems beyond mid-century and even fewer to the end of the century, a timeframe that energy, economy, and climate models often consider when assessing impacts of climate and technological changes. However, some technologies, including nuclear power and oil sands technologies, prompted much longer-term views on the environmental impact of the technologies' waste products.

The complexities of risks and uncertainties in transition pathways

The starting point for each low-carbon transition pathway were technologies, while the point of reference for the risk assessment were the policies (or policy choices) that enabled a certain pathway based on that technology. From the literature and initial stakeholder engagement, we found that two fundamental but related ideas about risks associated with low-carbon policies existed, which are part of scientific and public discourses in the context of climate change mitigation: risks as barriers to policy implementation and risks as consequences from implementing policies and supporting certain technologies. Each of the risks identified changes its character and meaning once we give up policy as the reference point for implementation and consequence: For example, a lack of investment, which may be a barrier to policy implementation, can in turn be a negative consequence of bad communication of policy elsewhere. Even more complicated, then, is the specification of subcategories, in our case political, policy, social, environmental, economic, and technological risks. These categories are common and feel intuitive, but they also are not clearly distinct from one another. Indeed, most risks fit several categories or cascade across these categories. For example, what is first only an economic problem, such as high upfront or maintenance costs, is something that is then passed on to consumers, affecting them differently based on their wealth, education, etc. – hence, it becomes an issue of social equality; or it is passed on to the environment, potentially increasing pollution in one place to reduce costs elsewhere. For our work, it turned out that in some instances the categories were well applied, whereas in others they were replaced with more fitting categories. Ultimately, categories are necessary to provide structure to the discussion, which is what their main purpose is, and can remain if we do not put higher hopes in them.

In the particular context of discussing decarbonisation policies, stakeholders tend to see implementation risks first and consequential risks only if prompted. Moreover, they frequently frame risk in economic terms, i.e. with respect to costs and investment needs. Environmental risks receive attention more because it has become an established category. Social aspects are relevant from an implementation perspective but are rarely considered in terms of consequences, potentially, because they are often a second- or third-order consequence rather than a direct result of a low-carbon transition. Chile and Indonesia were exceptions to this rule. Technological risks in terms of safety hazards were not explicitly touched on in most cases, even in the case of nuclear power.

Conclusions

In decarbonising their economies, countries of the world face a considerable challenge – even more so in light of the Paris Agreement and revelations of the 1.5°C IPCC Special Report. For this, they need low-carbon transition pathways which are based on: identified low-carbon technology options;

applicable good practice lessons from the past regarding scaling up these options; a supportive current national and international policy framework; and a deep understanding of national, regional, sectoral, and local contexts, including the priorities and responsibilities of stakeholders. Pathways for reducing emissions can impact other areas and aspects of society (consequential risks). Pathway implementation can also be hampered by political, market, or other types of system barriers (implementation risks). Narratives in this book describe the diversity of pathways and identify risks for each, and uncertainties where possible.

Including stakeholders in risk assessment is important as they, from their practitioners' perspective, are an important source of information for identifying risks associated with low-carbon transition pathways. While these risks are context-specific and reflect a broad range of stakeholder perspectives, common risks are found across the case studies, regardless of the local context, the technology promoted, and the policy mix implemented.

There are three nested dimensions present in all pathways which help both to explain the source of a risk and identify solutions to address these: *technological innovations, policy mixes*, and *society*. Each pathway is rooted in one or more *technological innovation* that often face cost barriers compared to existing technologies, as they often have not yet reached a stage of commercial applicability. Low-carbon technologies are promoted by *policy mixes*, including policy instruments to overcome financial barriers through reducing the higher cost risks of technologies. However, policy implementation in itself can be hampered by risks, too, such as insufficient *monitoring and enforcement*, particularly if co-ordination is required across different governance levels. Additionally, *policies* can be heavily driven by *politics*, the changes in which can create significant uncertainties and potentially dismantle existing policy support for low-carbon technologies.

Technological innovations and the associated policy mixes resonate in the *social dimension*, where their impacts are felt and support or opposition determined. Incumbent technologies are highly influenced by path-dependent *socio-economic interests* that enable them to exist, even if the technologies themselves present consequential risks linked to environmental impacts. *Societal acceptance* is also a crucial factor for scaling up technologies, especially when located near communities. Technologies that have a direct interface with end users require *behavioural changes* to ensure long-term sustainability of a pathways. The implementation of higher cost technologies can also have unintentional negative impacts on *social justice* linked to energy poverty.

Our narratives, drawn from a comprehensive research and engagement effort and summarised in Table 14.2, help to make explicit the barriers and negative outcomes that would have otherwise stayed implicit or hidden. Rather than addressing risks in an ad-hoc manner as they unfold, we can identify risks early on in the design of a transition pathway in an attempt to mitigate them before they manifest. The risk categories help us to consider risks in areas that we otherwise might overlook due to more pressing current socio-economic or political concerns, sometimes not directly linked to the transition pathway.

We also note that there are multiple priorities at stake when discussing different types of risk. We need to more clearly define: 'What criteria is being pushed through and for whom?' Additionally, stakeholders have uneven influence and power on the transition pathways: some have the ability to influence the direction of a pathway, while others have little or no say but are most adversely impacted. We urge a more transparent, inclusive, and comprehensive risk assessment in transition pathways with a wide range of stakeholders to make explicit the risks that are unique at the local level but with overarching issues that are relatable in other contexts.

Table 14.2 Key risks in each country study pathway

Country study	Implementation risk	Consequential risk
Pathways for incumbent large-scale technology systems		
Austria – iron and steel sector	*Technology (investment)* • Risk of poor timing for investment *Policies* • Lack of a co-ordinated national climate and energy strategy • Lack of evidence-based regulatory framework • Missing co-ordination of an integrated European policy *Socio-economic* • Risk of play-off between climate mitigation and social justice • Risk that the contribution of behavioural change is not included	*Technology impact* • Stability of grids and flexibility of energy system • Lock-in effects *Socio-Economic* • Households bear the main part of costs *Other* • Risk that resources are not considered (only energy)
Canada – Athabasca oil sands in Alberta	*Policies and politics* • Political barriers *Socio-economic impact* • Social rejection	*Technology impact* • Technology causes land-use change • Technology causes air and water pollution *Socio-economic* • Disruption of traditional ways of living • Negative health impacts • Drop in profits, contraction of the economy • Job losses

continued

Table 14.2 Continued

Country study	Implementation risk	Consequential risk
UK – nuclear power sector	*Technology (investments and others)* • High nuclear costs and limited profit margins • Lack of technology and specialised skills • Nuclear accidents *Policies and politics* • Conflicting policies and regulations *Socio-economic* • Social disruption and activism	*Technology impact* • Nuclear accidents/social disruption and activism *Socio-economic* • Radioactive waste legacy and intergenerational injustice

Pathways towards renewable electricity systems

Country study	Implementation risk	Consequential risk
Chile – renewable energy scale up	*Technology (investment)* • Renewable generated electricity costs *Socio-economic* • Opposition to diffusion of renewable energy projects • Quality of energy access across regions • Intra-country diversity in perception of energy needs *Other (external import factors)* • Dependency on fuel imports	*Socio-economic* • Energy poverty due to increased costs of electricity from energy impacting the budget of households • Energy vulnerability of households
The Netherlands – solar PV rooftop and parks	*Technology (investment)* • Lack of funding for solar parks *Socio-economic* • Hesitation to invest in solar PV *Other (environment)* • Lack of available land areas	*Technology impact* • Balancing issues power grid *Policies and politics* • High pressure on fiscal budget *Economic* • Energy company bankruptcies
Spain – renewable energy	*Technology (market)* • Oligopoly of the electricity market *Policies and politics* • Political instability • Volatile agenda of political parties • Regulatory changes	*Policies and politics* • Retroactive measures on renewables: subsidies cut • Discouragement of small actors and consumers • Competitiveness loss of renewables within the industry • Lack of incentives to finance the promotion of renewables

Table 14.2 Continued

Country study	Implementation risk	Consequential risk
Switzerland – renewable energy	*Policies* • Conflicting views on ES2050 • Complex permitting for infrastructure *Socio-economic* • Insufficient for public support for energy imports • Lack of acceptance of wind turbines • Principal agent problem for renters	*Technology impact* • Intermittent supply from PV • High cost of seasonal storage • Increased vulnerability of transmission grid to extreme weather for energy imports
Kenya – geothermal energy	*Technology (investments)* • High risk, high cost venture *Policies and politics* • Political uncertainty around national and county governance of resource use *Social-economic* • Community opposition to potential resettlement and changes in land access	*Technology impact* • Application of technology can lead to environmental degradation if process of monitoring and penalising is not clear and enforced *Policies and politics* • Tariff increases if supply–demand gap emerges • Lack of local benefit-sharing if Energy Bill continues to be delayed
Pathways energy efficient building sectors		
China – energy efficiency in the building sector	*Technology (investment)* • Lack of market financing channel/investment *Policies and politics* • Weak monitoring and policies innovation/enforcement • EE technologies innovation deviating the low-carbon aims *Social-economic* • Low perception of low-carbon transition and lock-in behaviours from the public	*Technological impact* • Technological improvements can lead to increasing land demands *Social-economic* • Social justice and climate/energy poverty • Increases in living cost

continued

Table 14.2 Continued

Country study	Implementation risk	Consequential risk
Greece – energy efficiency in public and private buildings	*Technology (investment)* • Lack of financial capacity to support implementation of technology *Policies and politics* • Political inertia; poor prioritisation • Bureaucratic complexity of energy efficiency-associated processes	*Policies and politics* • Tariff deficit overburdening the renewable energy sources fund *Social-economic* • Further impoverishing low-income households

Pathways using renewable household-level technology at the community level

Country study	Implementation risk	Consequential risk
Indonesia (Bali) – biogas in agriculture sector	*Technology (investment, supply chain and management)* • Initial investment needs • Accessibility and sustainability of feedstock • Poor maintenance and infrastructure • Collective management issues in the case of larger biogas systems *Policies and politics* • Inadequate monitoring practices • Lengthy and bureaucratic process to apply for governmental biogas programmes • Subsidy of fossil fuels • Lack of long-term incentives for biogas use	*Technology impact* • Unfiltered hydrogen sulphide (H_2S) in biogas installations • Methane leakage • Unsuitable technology for local conditions *Social-economic* • Shifting on household gender division of labour
Kenya – sustainable charcoal technologies	*Technology (investment and supply chain)* • High cost of more efficient production technologies and limited financing options • Competition from alternative clean cooking fuels • Limited capacity within charcoal producer associations *Policy and politics* • Weak enforcement of formal regulations	*Technology impact* • Efficient cooking technology can have possible environmental rebound effects from improved efficiency *Socio-economic* • Increased costs of sustainable charcoal and improved charcoal cookstoves • Potential loss of livelihoods within informal charcoal sector

Note

This table provides a summary of risks and does not make a distinction across various pathways identified in each country study.

Note

1 With specific insights from Alevgul Sorman, Eise Spijker, Oscar van Vliet, Rocio Alvarez-Tinoco, and Michael Stauffacher.

Reference

Unruh, G.C. (2000). Understanding carbon lock-in. *Energy Policy* 28, 817–830. https://doi.org/10.1016/S0301-4215(00)00070-7.

Afterword

Key insights on overarching risks across transition pathways

There is no 'one size fits all' low-carbon energy transition for all parts of the world. While this point is, perhaps, appreciated on a global scale, at a local level there are many lessons to learn. Economic, social, political, and environmental factors can vary enormously between different areas of a country, presenting a varied pattern of risks, opportunities, and challenges to the roll out of a low-carbon technology, associated policies, and behavioural changes. The narratives in this book reflect this variability in the diverse set of risks identified in each country. This said, we can broadly identify risks that are common across many, or all, of our case studies.

Investments in low-carbon technological innovations are often viewed as a high risk, which increases the cost of borrowing capital. Revenue guarantees, for example, can provide reassurance; however, national recognition and will to push forward with low-carbon technologies is not always backed up with the necessary funding to support it.

Small-scale technologies can encounter cost barriers at the household level, where the upfront cost of these technologies is born. Programmes to overcome these cost barriers are often implemented in only specific sub-national areas over a limited timeframe, which restricts the scaling up of a technology. Coherent, long-term policies are necessary to deploy these small-scale technologies at scale. At the same time this highlights the importance of action taken at lower administrative levels.

Many industrial-scale energy industries have large sunk costs in infrastructure and skills development. A low-carbon pathway that suggests abandoning or scaling back these industries can encounter significant resistance due to the destruction of financial and human capital. As a consequence, a feasible way forward can be through improvement of existing technologies which are unattractive from an environmental (including climate) perspective but nevertheless realistic from a social perspective.

Stakeholders expect low-carbon technologies to lead to positive environmental impacts, mainly through climate change mitigation or improved air quality. However, technologies can give rise to negative impacts, particularly when scaled up. These impacts are often linked to land-use changes but also include knock-on impacts in the production, installation, operation, and dismantling of a technology.

Government support to minimise financial risk through policy incentives to investors and households can in itself expose both governments and energy consumers to significant financial risks if the cost of the policy escalates. While governments can implement measures to manage this risk, they have limited control over many factors, such as the cost development of solar PV systems, electricity prices, and European energy and climate directives.

Heavy subsidies for low-carbon energy may have perverse impacts if the end users fail to achieve 'buy in' to a technology's deployment. When end users were provided with fully subsidised technology installations they often took an 'easy come, easy go' attitude to the technology and abandoned them, rather than attempting to rectify the problem.

Good governance is essential for successful implementation of low-carbon pathways. Scaling up these technologies often requires co-ordination across various government institutions at different governance levels. Co-operation can sometimes be difficult due to poor communication, differing priorities, and entrenched disagreements between different agencies. Good governance also applies to the policy support schemes. Monitoring and enforcement procedures are critical to a scheme's success but are often seen as minor components and poorly funded or implemented.

Dominant industries can be strongly entrenched within the socio-economic norms of society. Politicians are often hesitant to support low-carbon pathways if there are risks of disrupting the status quo, particularly if they think this will result in public backlash. Politicians who advocate a departure from the entrenched position can be great enablers of change but create high risks if political backing is withdrawn, particularly in countries with unstable political climates.

Public resistance to a low-carbon deployment can be reduced through careful design and stakeholder involvement. Public support is easier to achieve when technologies fit within the local landscape and do not have significant negative impacts on people. If public resistance is anticipated, early stakeholder involvement can reduce risks by allowing stakeholders to modify the scheme and achieve 'buy in'.

Successful deployment can also be improved by considering the environment and social-economic context of (sub)national regions. Areas within a country can vary hugely in their suitability for a technology and the economic ability of its residents to invest in a technology. Social practices can also influence successful technology deployment. We saw this in countries with diverse climatic regions which impacted programmes to improve the energy efficiency in homes.

Technologies and policies intended to reduce emissions may also unintentionally have a negative impact on social justice, including gender equality, and efforts should be made build a detailed understanding of their impacts.

Finally, the timescales required to mitigate climate change and the technologies necessary to combat it often fail to align with those of stakeholders. Households often judge impacts on their finances over very short timescales, while politicians are mindful of their next election. These short timescales contrast with popular framings of climate change and low-carbon technology, with models often looking to the mid-century and beyond when assessing economic and climate impacts.

Index

Page numbers in **bold** denote tables, those in *italics* denote figures.